GENTLEMEN ON
IMPERIAL SERVICE

1994

Sono Nis Press
VICTORIA, BRITISH COLUMBIA CANADA

# GENTLEMEN ON IMPERIAL SERVICE

## *A Story of the Trans-Pacific Telecommunications Cable*

TOLD IN THEIR OWN WORDS BY THOSE WHO SERVED

R. Bruce Scott

**Canadian Cataloguing in Publication Data**

Scott, R. Bruce
Gentlemen on Imperial service

Includes index.
ISBN 1-55039-053-8

1. Cables, Submarine—Pacific—History. I. Title.
TK5613.S36 1994 384'.1'09164 C94-910615-1

This book was published with the assistance of the Canada Council Block Grant Program.

Financially assisted by the Government of British Columbia through the B.C. Heritage Trust.

Published by
SONO NIS PRESS
1745 Blanshard Street
Victoria, British Columbia
Canada V8W 2J8

Printed and bound in Canada by
MORRISS PRINTING COMPANY LTD.
Victoria, British Columbia

DEDICATED TO FORMER

*Pacific Cable Board Veterans*

—PAST AND PRESENT

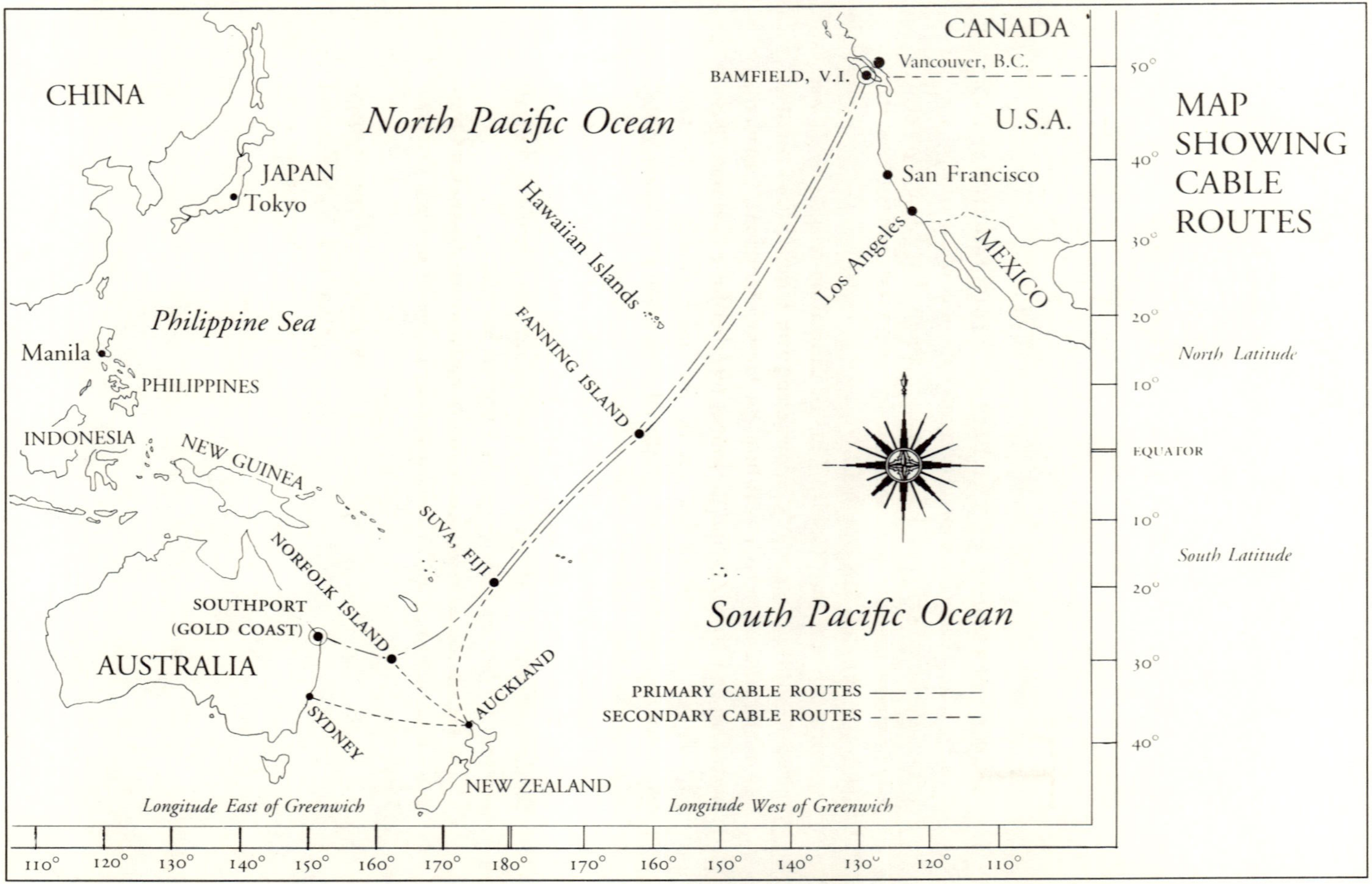
MAP
SHOWING
CABLE
ROUTES
CANADA
Vancouver, B.C.
BAMFIELD, V.I.
U.S.A.
San Francisco
Los Angeles
MEXICO
CHINA
JAPAN
Tokyo
North Pacific Ocean
Hawaiian Islands
FANNING ISLAND
Philippine Sea
Manila
PHILIPPINES
INDONESIA
NEW GUINEA
SUVA, FIJI
NORFOLK ISLAND
SOUTHPORT
(GOLD COAST)
AUSTRALIA
SYDNEY
AUCKLAND
NEW ZEALAND
South Pacific Ocean
PRIMARY CABLE ROUTES
SECONDARY CABLE ROUTES
Longitude East of Greenwich
Longitude West of Greenwich
North Latitude
EQUATOR
South Latitude
50°
40°
30°
20°
10°
10°
20°
30°
40°
110°
120°
130°
140°
150°
160°
170°
180°
170°
160°
150°
140°
130°
120°
110°

# *The Pacific Cable Board*

THIS IMPRESSION OF THE PACIFIC CABLE BOARD'S BADGE or insignia may bring back many pleasant memories of the years of duty spent at the Board's city and Pacific Island stations. The P.C.B. operated only between the years 1902 and 1929. In 1929 it was absorbed into the Cable and Wireless Ltd., system.

Maybe this was the "Golden Era" of submarine telegraphy as we knew it—with simplex and duplex working over a single line—twenty-four hours a day. A service that brought United Kingdom, Canadian, Australian and New Zealand personnel together for a common cause. Capable management, skilled electricians and engineers, competent operators and rigid discipline produced a standard of efficiency and accuracy that has seldom been equalled.

In Canada, in Australia and in New Zealand, surviving members of the P.C.B. meet annually to renew old friendships and to reminisce over the days when the Directors of the Board in London referred to the staff overseas as "British Gentlemen on Imperial Service."

The logo depicted above has been presented to the Australian Overseas Telecommunications Commission for inclusion in their Veterans' Museum.

W. CRAIG—*former manager*<br>
Auckland, New Zealand<br>
September 1980

# *Acknowledgements*

WITH SPECIAL THANKS TO KIMBERLY O'SULLIVAN OF THE Australian Overseas Telecommunications Corporation's Archives Department for her valuable assistance, and to Barbara Pell for restoring my confidence with her encouragement, to John Wallace, Alfred Gallagher and John Langmead for checking the manuscript, to Paula Johanson for her criticism and suggestions, and to Angela Addison for her valuable advice.

Portions of chapters four and five appear in slightly different forms in *Barkley Sound*, a history of the Pacific Rim National Park area.

# *Preface*

THIS IS THE STORY OF THE TRANS-PACIFIC TELECOMMUNICATIONS cable, a link in a chain of submarine cables connecting Australia, New Zealand, Fiji and Canada to the United Kingdom. Containing historical, technical and human interest material, it outlines the development of the system from simple manually operated instruments to modern technological improvements. It relates the experiences of members of the staff who served in isolated stations in the South Seas, Australia, New Zealand, Fiji and Canada. Their services are typical of those provided by other cablemen.

It covers the Pacific Cable Board era through subsequent administrations of Imperial and International Communications, Cable and Wireless Ltd., Australian Overseas Telecommunications Commission, Canadian Overseas Telecommunications Corporation and Teleglobe, through World War I, the influenza epidemic of 1918 to modern times.

It also provides a behind-the-scenes view of what was involved in the transmissions and reception of overseas messages, and of the lifestyle of that era.

Having spent my whole working life, 1920-1960, in the trans-Pacific Cable service, and as one of the few surviving members of the Pacific Cable Board, I feel that it is incumbent on me, in the latter years of my life, to write what I know, and have learned through research, about this important, strategic project—especially since its story has never been told before.

Previous Books by R. Bruce Scott

*Breakers Ahead!*
A History of Shipwrecks on the Southwest Coast

*Barkley Sound*
A History of the Pacific Rim National Park Area

*People of the Southwest Coast of Vancouver Island*
A History of the Southwest Coast

*Bamfield Years: Recollections*
A Personal Experience of Life in a Picturesque Fishing Village

# *Contents*

CHAPTER
ONE

# Joining the Pacific Cable Board

AT THE END OF WORLD WAR I, I was a teenager earning fifteen shillings a week as an office boy in Sydney, Australia. I remember the celebrations at the end of hostilities in 1918. People danced in the streets. Everywhere there was rejoicing. However the celebrations did not last long. They soon turned to grief and sorrow as thousands of people died from the influenza epidemic. Unable to cope with the enormity of the problem, authorities stacked bodies on the railway station at Blackwood, where the cemetery was located, to await burial. Everyone wore masks of cotton gauze which was supposed to prevent infection.

My father, then in his forties, had enlisted in the war because he did not want my elder brother to do so. Father was a Chartered Accountant and an accomplished pianist who often accompanied prominent artists performing in the Sydney Townhall. In earlier years he had been organist at St. Peter's Cathedral in Adelaide.

Shortly after returning from the war, he died as a result of a combination of "shell shock" and erisipelas, leaving mother to raise five children on a paltry widow's pension. She supplemented this by taking up hairdressing, making her own lotions and treating her clients in their own homes.

I had to leave school at the age of thirteen and go to work as an office boy. I moved from job to job, but my heart was not in any of them. One was as a counter clerk in the button department of Hordern Brothers, a large department store. Why they placed me in that position I do not know. I did not last long. I continued to look in the Help Wanted advertisements in the *Sydney Morning Herald.* One day my eye caught the following:

> WANTED, boys fifteen years of age to learn submarine telegraphy and serve overseas. Apply Pacific Cable Board, 77 Pitt Street.

Intrigued, I asked my Boy Scout friend, Bob Addison, who was an apprentice in the Amalgamated Wireless Company (later to become the Beam Wireless Company), what he thought about it.

"Oh! That's the latest way of communicating between submarines," he said. Intrigued even more, I decided to apply for the job with my mother's approval. I soon learned that submarine telegraphy meant underwater cable telegraphy and had nothing to do with submarines. If accepted, I would serve as office boy in Sydney for about twelve months before being transferred to the Southport Cable Station in Queensland, where I would be trained as a cable operator.

I was accepted and began work in the old 77 Pitt Street office (since demolished) of the Pacific Cable Board.

There were always about four boys making up "carbons" while awaiting their turn to go to the training school. It was a monotonous job. It consisted of inserting carbon paper into the various message forms—thousands of which were used every day. It was not at all like the romantic occupation I had imagined.

I stood it for about two weeks then, one Saturday afternoon, I watched from the office window the crowds of people streaming down Pitt Street towards the Sydney Harbour ferries and an afternoon on the beach. That was the last straw! I decided to quit the job. Everyone else had Saturday afternoons off while we had to work eight hours a day, six days a week!

The following Monday I went upstairs to the manager's office to resign, but he was out. I went back several times that day but he was out each time. By the next day I had cooled off and decided to "give it another go."

On that slender thread lay my whole future. I remained in the service for the rest of my working life—forty years—and never regretted it.

When the soldiers and sailors began returning from the war, a Voluntary Workers' Association was formed by the public to build homes for them on land in French's forest, on the Manly side of the Spit Bridge. Every weekend truckload after truckload of voluntary workers consisting of all kinds of tradesmen and professionals, and their wives, barrelled down the highway over the Spit Bridge, singing familiar songs such as, "A Long, Long Way to Tipperary," "Mademoiselle from Armentiers," etc.

I shall always be grateful to the Mosman Bowling Club, of which my father used to be a member, for building a house for my widowed mother and her five children on Beauty Point, near the Spit Bridge, in Mosman, by means of the voluntary labour of its membership, which included architects, doctors, accountants and other professionals.

CHAPTER
TWO

# Southport Training School

WHEN THE TIME CAME TO BE transferred to Southport, I was issued with a memorandum advising the time, date, and point of departure on a local steamer for Brisbane, the capital of Queensland. As parting gifts, my mother gave me a steamer rug and a dictionary—both of which I still have, although the latter is rather tattered and torn. I read it from cover to cover many times, making note of the words that appealed to me.

At Brisbane I boarded a train for Southport, fifty miles away. Upon arrival I was met by a horse-drawn carriage and driven to the Cable Station on Bauer Street. Southport had an endless stretch of beautiful sandy beaches completely devoid of any development. The Nerang River ran parallel to the beach, separating the town from the beaches. To get to the beach, we crossed the river on a small passenger ferry run by Barney Bolton, who also ran a small store nearby at which the cable boys were able to buy cigarettes, non-alcoholic drinks and candy on "tick."

Near the centre of town there was a wide pier jutting out into the sea where the river emptied. At one end of the pier there were an enclosed swimming bath and a band rotunda.

The Cable Station and Training School was located close to town. Like all cable stations it had a tennis court, gymnasium and library. There was a fair amount of social activity in the town of about five thousand people, to which cable boys were usually invited. For social occasions the boys wore navy blue blazers trimmed with red embroidery, with a red crown emblazoned on the left hand breast pocket: the insignia of Imperial Service.

Upon arrival at the school a boy was shown to his bedroom which was equipped with an iron bed, wash-basin and jug of cold water, a writing desk and chair. The accommodation was under the management of a Matron and her husband, who also served as cook and waiter. There were always a dozen boys resident in the school.

Instruction in cable operating began the next day. It included reading long distance cable signals from various stations in the Pacific area. The cable between Bamfield and Fanning Island was the longest in the world—over four thousand miles—so these were the most difficult to read. Cable signals, although in Morse Code, were not like telegraph signals. They consisted of a wavy line like a seismograph record, on a narrow tape with dots above and dashes below the centre.

Other modes of telegraphy were: sending and receiving the Morse Code with key and brass sounder, typing up ten-letter code words, transmitting messages by hand-perforator and keyboard perforator, transmitting by cable keys—which were different from telegraph keys in that they had two keys—one for dots and one for dashes—thus reversing the current in the cable. In all cases emphasis was placed on cultivating the correct wrist action for tireless operation hour after hour.

To quote the P.C.B. rule book, "The course of instruction should be such as will bring about the power of youths to key, perforate and type on the several instruments in a correct manner free from blundering, and to receive with accuracy." It was necessary, said the rule book, to stamp out the erasure habit and avoid carelessness of any description.

The school proved to be a challenge to my ability. I devoted myself wholeheartedly to the challenge and felt that at last I had found my niche in life. Some boys were unable to adapt to the work and were transferred back to Sydney on clerical positions.

After a boy reached a certain degree of proficiency, he was assigned to duty in the operating room for a few hours each day and sometimes alone on night duty. There was little or no traffic routed via Southport at that time. The main route was Sydney-Auckland-Suva-Fanning Island-Bamfield, with the Southport-Norfolk Island-Suva section forming an alternative emergency route.

The main job of the man on night duty was to send extracts from the Brisbane daily newspaper by key over the cable to Norfolk Island for the local "newspaper." These were typed up by the Norfolk Island operator and tacked up on a tree at the crossroads for all to read. The tree was referred to as the "Tree of Knowledge" by the islanders.

One night while I was alone on night duty, Norfolk Island began sending an urgent S.O.S. message from the sailing ship *France*, which at that time was the largest sailing vessel afloat. She had gone ashore in the New Hebrides. As Norfolk Island was unable to make radio contact with Brisbane, the message was sent to Southport over the cable for retransmission by Morse telegraph to Brisbane. For a junior operator, it was an exciting moment! So much depended on me—possibly the safety of the crew. I managed to control my nervousness. Calling Brisbane on the Morse key, I relayed the message which was acknowledged. Unfortunately the ship broke up before help could arrive.

Examinations to test the proficiency of the boys were held at the end of each month. Boys had to pass the test three months in succession before being eligible for transfer back to Sydney as probationary operators. Instruction could take anywhere from nine months to two years, depending on the aptitude of the boy. A small salary was paid during tuition.

The main attraction at Southport was the ocean-beach surfing. The beach, now known as Surfers' Paradise or the Gold Coast, consisted of miles of sandy beach completely devoid of development. Now there are high-rises from one end to the other. Not many townspeople participated in surfing in those days, but the cable boys, being familiar with the system of voluntary lifeguards on most Australian beaches, decided to form their own surf club to patrol the beach and render assistance to those caught in the undertow which, at times, could sweep the unwary out to sea. During weekends, hours were assigned to each boy for beach patrol. Not being an enthusiastic surfer, I did not show up on one occasion and was roundly denounced. So much so that it ended up by my being challenged to fight by Sam Nathan, the biggest member of the group. Although never having fought before, except in early boyhood days at the Neutral Bay school, I felt bound to accept. It did not last long. I was knocked out in the first round. That cleaned the slate and I was accepted again.

One day another cable boy, Alf Kelso, and I decided to go horseback riding. We rented two horses from the local livery stable, which supplied horses for the carriage trade, and went exploring tree-lined country roads for several hours. We returned to the centre of town and galloped down the main street like two outlaws with the sheriff's posse in hot pursuit. Although it was the first time I had ever been horseback riding, I did not feel any after-effects.

After passing the examinations in the first year, I was transferred back to Sydney as a probationary operator. I lived there with my mother, and sisters Thelma, Patricia and Janet, my brother Harold having left home and married.

CHAPTER THREE

# The Trans-Pacific Cable

THE HISTORY OF THE TRANS-PACIFIC submarine telecommunications cable illustrates the immense technological progress over the past century in overseas telecommunications. Beginning with a primitive system manually operated at a speed of 120 letters per minute it has now developed into one that can handle 60,000 simultaneous telephone conversations between Sydney and Auckland as well as additional services such as video-phones, colour facsimile, high resolution television and other forms of high-speed information.

In 1873 a cable service linking the United Kingdom with Australia was established via the Middle East, called Via Eastern. But an additional route was also required to connect countries in the Pacific area with the United Kingdom, thus completing an around-the-world system of communications. It was to land only on Empire territory providing a reliable, secret means of communications that could be defended in time of war.

Conceived by Sir Sandford Fleming, Chief Engineer of the Canadian Pacific Railway Company, the proposal was discussed for over a quarter of a century, the main subject of discussion being the route of the cable. The proposed section between Bamfield, a small isolated fishing village on the west coast of Vancouver Island, and Fanning Island, a tiny coral atoll in mid-Pacific, was over four thousand miles long—far longer than any other submarine cable in the world.

"The electrical capacitance would be too great," experts claimed. "No signals would come out at the other end," they said.

At the request of the Canadian, Australian and New Zealand governments, the British government eventually agreed to the formation of the Pacific Cable Board, which consisted of eight members, three British, two Canadian, two Australian and one New Zealand. The Board was instructed to agree upon a route, then call for tenders and report the results to the governments concerned, who would then have to decide unanimously to proceed along lines recommended.

Many routes were suggested and discarded. The one finally agreed upon was Bamfield-Fanning Island-Suva-Norfolk Island-Southport, then overland to Sydney, with a branch line from Norfolk Island to Doubtless Bay, in New Zealand, then overland to Auckland. It was to have a guaranteed speed of no less than twelve words a minute (sixty letters).

Once the route was agreed upon, the project was undertaken by the Imperial (British) and Empire governments under the aegis of the Pacific Cable Board.

The contract for making and laying the 8,000 miles of cable was awarded to the Telegraph Construction and Maintenance Company at the price of 1,795,000 pounds sterling, which at that time of the gold standard, was worth $8,723,700 Canadian. A loan of two million pounds at three percent interest was obtained from the British government.

Actually the experts who criticised the project were right to a certain extent. During final tests, after the cable had been laid, engineers found that it took 9.3 seconds for a single signal from Bamfield to reach Fanning Island! The letter S takes three signals (three dots) so that only two letter S's could be sent in one minute. However the engineers conceived a method of dealing with this problem known as "signal shaping." The result was that the cable was worked at 120 letters per minute—twice the guaranteed speed—which was an easy speed for the operators to maintain, considering the difficulty of reading the long-distance cable signals.

CHAPTER
FOUR

# Surveying the Route

THE LAYING OF THE LONG CABLE WAS a great feat of engineering. Before laying the cable, it was necessary to ascertain not only the various depths, but also the nature and contour of the bottom of the ocean along the route so that when the cable was being laid, the requisite amount could then be paid out to follow the contours and avoid draping the cable from peak to peak of underwater mounts.

The Telegraph Construction and Maintenance Company's cableship *Brittania* began surveying the deepsea route on 30th August, 1901.

In March, 1901, the Canadian government ship *Quadra* left Victoria, B.C. in search of a suitable landing site. She proceeded to Jordan River where the foreshore was examined. From there she went to Port San Juan and, after examining that locality, steamed into Barkley Sound. Here the most favourable site was found at Bamfield where the sea bottom consisted of black ooze devoid of rocks. The gradual slope of the shore was another desirable feature. Other factors that influenced the choice of the site were: the distance from man-made electrical disturbances which would have distorted signals; less danger of the cable being fouled by anchors than in a busy port like Victoria or Vancouver; every additional mile would have added resistance and capacitance and reduced the working speed of the cable.

In July, 1901, HMS *Egeria* based in Esquimalt, B.C. commenced taking soundings on the west coast of the island under the observant eyes of W. P. Daykin, lightkeeper of the Carmanah Point lighthouse. One of the outstanding characters of the southwest coast of Vancouver Island in the early days, Daykin kept official lighthouse diaries consisting of eleven volumes in which he recorded every ship sighted from the vantage point of the lighthouse tower, and every happening of interest up and down the coast. On this occasion he noted:

| | |
|---|---|
| *12 July 1901* | British man-of-war two miles SW off here all afternoon with her boom strung out. I hoisted our ensign and saluted. She answered. |
| *13 July 1901* | The man-of-war still to westward. She seems to be taking soundings. |
| *14 July 1901* | Man-of-war still at anchor in same place. Had several boats down. |
| *15 July 1901* | Man-of-war *Egeria* still west. |
| *17 July 1901* | The *Egeria* still surveying off Barkley Sound. 6pm *Egeria* sounding off Bonilla Point. |
| *21 July 1901* | *Egeria* in San Juan |
| *22 July 1901* | *Egeria* heading west. |
| *27 July 1901* | *Egeria* still surveying. |
| *29 July 1901* | *Egeria* passing Cape Flattery inbound. |
| *24 August 1901* | *Egeria* outbound. |

On October 9, 1901, the Victoria *Colonist* newspaper reported that:

> A telegraph construction party left Victoria on the coastal steamer *Queen City* bound for Bamfield. It consisted of: J. Wilson, superintendent of Canadian Pacific Telegraphs, Mr. Cambie, chief engineer of C.P. Telegraphs, Mr. Cartwright,—assistant engineer of the Lulu Island Railway and Mr. Lockwood who was in charge of the arrangements. The actual construction would be done by workmen who would be sent as soon as these preliminary arrangements had been completed.

## South Pacific Cables

Next year the cableship *Anglia* laid the south Pacific cables from south to north in the following order:

| | |
|---|---|
| *8-18 March 1902* | Southport-Norfolk Island 837 n.m. |
| *20-25 March 1902* | Doubtless Bay-Norfolk Island 519 n.m. |
| *3-10 April 1902* | Norfolk Island-Suva 980 n.m. |

There was a delay of five months from April until September while a special cableship, *Colonia*, was constructed to load and lay the 3,459 nautical mile (4,035 land miles) cable between Bamfield and Fanning Island. This North Pacific section of the cable was laid by the *Colonia* between 18 September and 6 October, 1902.

The section between Fanning Island and Suva was then laid by the *Anglia* between 17-31 October 1902. It was the last to be laid, completing the contract two months ahead of schedule.

In the spring of 1902, the observant Daykin made the following entries in his diary:

| | |
|---|---|
| *21 April 1902* | *Queen City* arrived southbound with a large gang of men going to the Cable Station in Bamfield. |
| *15 September 1902* | Alberni and Bamfield (landline) wires within four miles of each other. |
| | *Colonia* leaves Victoria on 17th September for cable station (to commence laying Pacific cable). |
| *17 September 1902* | *Colonia* passing outbound. Signalled her but received no answer. |
| *18 September 1902* | Cable ship *Colonia* left Bamfield at 3.30pm (commenced paying out cable). |

## North Pacific Cable

The *Colonia* arrived in Victoria on 12 September, 1902, with over 3,459 nautical miles of cable on board and proceeded to the naval base at Esquimalt, B.C. for coal and supplies before sailing for Bamfield. After having loaded 1,600 tons of coal, she sailed at noon on the 17th September for Bamfield to commence laying the long cable.

At Bamfield she stood off some four hundred yards from shore and prepared to lay the heavily armored shore end. An endless rope, like a continuous clothesline, was rigged between ship and shore. A large swivel block, made fast to a tree stump, served to anchor the rope which conducted the end of the cable to the cable-hut ashore. Empty barrels attached at intervals kept the cable afloat until the end was safely ashore, when they were cut adrift allowing the cable to sink to the bottom.

The late James McCay, ex-Bamfield pioneer, was present when the cable shore-end was landed. "I was just a small boy," he said, "but I remember one of the sailors saying to me, 'Come on boy, grab hold of this line.'"

At two o'clock in the afternoon, with a parting salute to the officials ashore, the *Colonia* began paying out the cable on her voyage to Fanning Island four thousand miles away. As was customary on such occasions, local residents who had gathered to watch the historical event were served refreshments in the traditional manner by Mr. McLachlan, superintendent of the Cable Station. The *Colonia* laid the long cable in just under eighteen days at an average speed of eight knots—which was a record. It was the longest cable ever laid in one piece.

In the meantime construction of the main building in Bamfield was nearing completion. It was designed by Francis Mawson Rattenbury, the same architect who designed the Parliament Buildings and the Empress Hotel in Victoria.

Clearing the site of huge cedar trees had been a problem in itself, the efforts of the workmen being limited to falling the trees and cutting them into cordwood for use in the station furnace. The stumps were left in the ground, and in many places, are still there. One or two of them came in handy for mooring the large cableships, the wharf being too small for that purpose.

The building, a large photogenic three-storied structure resembling a Canadian Pacific hotel, or French chateau, was situated on a hilltop commanding panoramic views of Barkley Sound. It would provide not only office and maintenance rooms, but also accommodation for the twenty bachelors on the staff. It contained about fifty rooms including a large dining room with view windows, billiard room, library with three thousand books and an open fireplace, music room, etc. Accommodation was provided in the basement for a dozen Chinese men, half of whom would act as servants while the remainder would be maintenance men of the property. There were also twelve houses for married men.

The whole project was finally completed on 1st November, 1902.

CHAPTER
FIVE

# Opening Celebrations

THE PERFORMANCE OF THE CABLE exceeded the design engineers' expectations. Despite an electrical resistance of 6,295 ohms and an electrical capacitance of 1,450 microfarads, a duplex speed of 110 letters per minute was accomplished — almost twice the guaranteed speed. This speed was gradually increased to 135 letters per minute by improved methods of signal shaping.

A special celebration to mark the completion of the cable was held in the old Victoria Theatre. The stage was decorated with a large map showing the "All Red Route." In those days the British Empire was coloured red on all maps of the world, and since the cable landed only on Empire territory, it was called the "All Red Route." A table containing telegraphic apparatus was placed on the stage near the footlights. Acting as operators were Mr. Christie, manager of the local Canadian Pacific office, and Robert Bain, assistant superintendent of the Bamfield Cable Station. Congratulatory messages were exchanged with various heads of state and messages from all parts of the world poured in to Sandford Fleming congratulating him on the success of his great project.

Mayor Hayward expressed the feelings of the gathering when he said, "It could not fail to be a source of pride and satisfaction to every member of the British Empire that this important enterprise had succeeded. It will form a bond of brotherhood and prove of great benefit to every part of the Empire."

Following the completion of the trans-Pacific cable, winter storms soon showed up the weakest link in the around-the-world system. This was the tenuous thread of copper wire strung from tree to tree along the eastern shore of the Alberni Inlet. Gale force winds frequently blew down numerous trees across the line, interrupting the whole service between Sydney and London for weeks at a time. Linemen had the laborious task of cutting away the trees with only crosscut saws and axes, sometimes during heavy snowfalls. From Alberni the line went by an overhead line strung on telephone poles via Cameron Lake to Parksville, then underwater by cable to Vancouver.

The following summer the new cableship *Iris* arrived in Esquimalt with 45 miles of cable for the Bamfield-Alberni section, replacing the primitive landline. Right from the beginning some experts had claimed that it would have been better to have run the cable into the vicinity of Vancouver instead of Bamfield. Fifty-seven years later the cable was entrenched between Port Alberni and Parksville, replacing the overhead landline.

For the first few years the Pacific cable was operated as if it were a private line, between various overseas government departments. As a consequence it performed at a loss. When placed on a commercial basis it was so profitable that it paid for the duplication of the Suva-Bamfield section in 1926 at a cost of $850,000.

The first extension to the system was made in 1912 when the cableship *Iris* diverted the Norfolk Island-Doubtless Bay cable into Auckland, and the *Silvertown* laid a 1,225 nautical mile cable between Sydney and Auckland. Further extensions were made in 1923 when the cableship *Stephan* laid a cable between Suva and Auckland. She then laid a cable between Sydney and Southport, bypassing the overland telegraph line.

A surprising amount of messages could be transmitted over a slow cable when it was kept full twenty-four hours a day. At the prevailing duplex speed of 24 words (120 letters) a minute, 34,560 words could be moved daily in each direction. Additional space could be saved by using abbreviations. At Christmas time the text of messages containing "Merry Christmas Happy New Year" was abbreviated to MXHNY which the receiving operator typed up in full, thus increasing the capacity of the cable and the speed at which the operator had to work. It was considered merely part of the day's work.

CHAPTER
SIX

# Methods of Operating Submarine Cables

WHEN THE FIRST CABLE WAS LAID across the Atlantic Ocean in 1857, messages were received in a very primitive manner compared with modern electronic methods of instant printouts. An instrument containing a coil of fine copper wire suspended between two magnets was used as a receiver. A small mirror was attached to the coil and a beam of light was centred on the mirror and reflected back on to a horizontal scale. This magnified the weak signals which were then read by an operator in a darkened room observing the fluctuations on the scale. The light moved to the left for a dot and to the right for a dash, remaining in the centre for zero. The receiving operator called out the letters to another man who wrote them down at a speed of approximately eight words a minute. Later it was found practicable to project the moving light on to the writer's pad, thus eliminating the reader and avoiding sound, or phonetic errors. A disadvantage of this system was the lack of a permanent record that could be consulted a second time in the case of doubtful signals.

Cable signals were different from telegraph signals. Instead of making and breaking a contact by means of a telegraph key, the electrical current in the cable was reversed by means of a double key—one for dots and one for dashes. Both dots and dashes were of the same duration, unlike telegraphy where dashes were three times the length of dots. The Morse code was used in both cases.

The next improvement was the siphon recorder. Instead of a mirror, a thin glass tube was attached to the galvanometer coil which siphoned ink out of a small inkwell above the coil. Incoming signals moved the lower end of the siphon from side to side, marking the signals on a moving tape, much like a seismograph. When there were no incoming signals, the siphon marked a line in the centre of the tape. It marked above the centre line for a dot and below for a dash. This tape was then fed through a glass tube placed above the keyboard of a typewriter. The tube contained water which magnified the signals. The receiving operator typed up the messages on appropriate forms. Sitting beside him was another operator who checked the work of the receiving operator for accuracy.

For transmission, a twin key was used to reverse the current in the cable, the positive current marking a dot and the negative a dash. This system of manual transmission by key was used for many years, some operators attaining an amazing degree of dexterity, producing signals with machine-like precision.

The key was superseded by manually operated hand punches which perforated an oiled tape which was then fed into an automatic transmitter. These punches had three tappets, one for dot, one for space, and one for dash. The operator hit the tappets with two rubber-tipped metal punch sticks, or mallets, in proper sequence. A high speed could be attained by experts who could play well-known tunes while operating—even cross over hands without missing a beat.

After hand perforators came the Kleinschmidt and Creed perforators, each with a keyboard and operated in much the same manner as a typewriter, except that they perforated a tape instead of printing on paper. They were used on all stations except automatic relay stations.

The reading of long distance cable signals called for a high degree of skill. The longer the cable, the worse the signals. The signals, as recorded on tape, had little definition and took an experienced operator to distinguish between the various combinations of dots and dashes. In addition, disturbances on the cable produced distortions and added to the difficulty of reading the signals. A long cable, bisecting as it does the lines of force of the earth, acts as a generator of spurious signals and any movement of these lines of force, or any lightning, generated a current in the cable and distorted the signals, some of which had to be repeated. In later years, however, most of the spurious signals were filtered out and bypassed.

To experienced operators, reading signals on the cables was almost as easy as reading manuscript. On a short cable the signals were sharp and clear, but on a long cable (such as the original cable between Fanning Island and Bamfield) the operator would have to judge the length of a signal rather than the number of dots and dashes.

For many years the "gum and punch" method was used at relay points wherever the cable landed. The tape containing the received signals was gummed down on a paper and then passed over to the transmitting operator who "punched" the signals on a tape which was then fed into an automatic transmitter. The operators were, in fact, "human" relays.

Many messages sent over the cables were in code and cipher, which increased the difficulty of reading the signals. An operator could have a good idea of what a letter in a plain language word should be, even though it might be distorted, but in code, he could not guess, he had to be sure. Every error was traced to, and logged against, the man who made it. Operators were allowed only a certain percentage of errors in the number of words they handled. If they exceeded that percentage in the course of a year, they lost their annual increment and seniority. It was a hard school, but it produced good operators.

During the Pacific Cable Board era, monthly error percentage lists were posted showing how many errors each operator had made based on the number of words transmitted and received. The operators strove to be at the head of the list.

During the Cable and Wireless era annual service certificates were issued showing the level of proficiency obtained by each man. In 1934 my certificate read:

> An extremely accurate operator and a good all-round man.
> For 12 months in 1932, 544,040 words, errors 2.
> For 12 months in 1933, 510,060 words, errors nil.
> For 4 months in 1934, words not taken, 1 error.
>
> SIGNED *W. R. Rutherford,*
> Manager

I salvaged this certificate from my personal file after having been instructed to destroy all staff records by the new Corporation that took over the service from Cable and Wireless Ltd.

The error percentage list was discontinued after this because the union, which the staff had just joined, disapproved of operators being penalized for making errors.

CHAPTER
SEVEN

# Repairing Deepsea Cables

THE FIRST TRANS-PACIFIC SUBMARINE cable was a telegraphic one like all others at that stage of the art. It consisted of a copper wire core about one eighth of an inch in diameter, covered by an insulating layer of gutta-percha (a rubber-like substance), then a strip of brass tape wound spirally to protect the core from teredos (marine worms), then jute packing and an armor of steel wires. The deep-sea cable was about one inch in diameter, but the shore ends, being heavily armored with steel wires to protect them from wear and tear in shallow waters, were about two inches in diameter. They also contained an additional copper grounding wire which extended out to sea for several miles to neutralize local electrical disturbances.

There is almost no tidal motion on the ocean floor where the deep-sea cable lies. In one place on the Bamfield-Fanning Island section, the cable rests at a depth of three and a half miles. It is rarely broken except by submarine earthquakes or earth-slides. Most breaks occur in shallow waters due to tidal movements chafing the cable on rocks, or ships' anchors hooking onto and damaging the cable.

Should an interruption occur, due either to a complete or partial break, the fault can be located from either end by measuring the resistance and capacitance of the cable to the point where the signals leak to earth. As the resistance and capacitance of the cable per nautical mile are known factors, the position of the fault, which is the unknown factor, can be calculated from the readings obtained by the Wheatstone bridge test. The nearest cableship would then be dispatched to repair the cable.

Sometimes a small marine animalcule may bore through the insulation of the cable, exposing the core to sea-water. A temporary repair can be made by applying a sealing current from the closest cable station, which causes a chemical coating to form on the exposed core, insulating it from the sea. However, before long the cableship has to go out and repair the faulty section. Arriving on location, the cableship lowers its grappling hook and criss-crosses the route of the cable making "drives" until the cable is hooked.

Despite the perfection of modern technology a primitive process is still used to detect the moment when the cable is hooked. An officer sits on the taut grapnel rope where it passes over the bow sheave and, by the feel of the vibrations through his posterior, can tell the nature of the ground over which the grapnel hook is passing and can feel the change in tension when the cable is hooked. Once hooked the good end of the cable is cut and carefully hauled to the surface, tested, sealed and buoyed. The ship then proceeds to the other side of the fault and repeats the process. A new section of cable is spliced in between the buoyed ends, tested and then released.

This type of telegraph cable has been superseded by modern telephonic cables of greater flexibility and capacity. However the method of recovery and repair is still much the same.

An illustration of modern methods of repairing submarine cables occurred in 1977 when communications on the Guam-Madang section of SEACOM telephone cable in the north Pacific Ocean were interrupted. Tests located the fault in the vicinity of repeater R8, about 160 km from Guam. Repeaters, like amplifiers, were built into the cable every twenty miles or so to amplify the signals and increase the bandwidth of the carrier, allowing more channels to be utilized.

The fault was located in the abysmal depths of the Mariana Trench in the Philippine Sea, one of the deepest ocean canyons in the world—nearly six miles deep!

The cableship *Edward Wilshaw*, commanded by Captain Watts, arrived on location and began grappling for the cable on the south side of repeater R7. After three days of "fishing" the cable was hooked at a depth of 2,500 fathoms (15,000 feet) and brought to the surface. It was then cut, sealed and buoyed. The ship then moved to the Guam side of repeater R8. After three days of grappling, the cable was hooked and was being hauled to the surface when it parted and fell back to the bottom of the ocean. The ship then went to the Madang side of repeater R8 and commenced grappling at a depth of 3,500 fathoms (21,000 feet). The cable was again hooked but parted again during recovery. Somewhat disheartened, the ship returned to port to re-fuel and replenish supplies.

Upon returning to the cable grounds the ship continued grappling operations, but after eight "drives" there was no sign of the broken cable. The search for repeater R8 was therefore abandoned and grappling for the Guam end of repeater R9 was commenced.

The cable was hooked at 4,500 fathoms (over five miles deep) which is the deepest cable recovery ever made. A new section of the cable was spliced to the old and relaid. The ship returned with the new end to the location of the R7 buoy to splice the two cables together, only to find that the buoy had disappeared!

Grappling for the cable began all over again, and fortunately, it was retrieved a day later. Both ends were brought on board, spliced and repositioned in the sea. However when tests were conducted by the terminal stations they were unsatisfactory. Transmission levels were faulty and the system too noisy.

There was nothing for it but to recover the cable once more, for the sixth time. However this recovery was simple, since the crew had taken the precaution of tying a rope to the cable, thus simplifying its recovery.

The fault was found, cut out, the final splice made, and the cable lowered to the bottom again. This time the tests proved satisfactory.

The whole operation had taken forty days.

In October, 1929, an immense submarine earthquake and subsequent landslide broke sixteen of the twenty trans-Atlantic telegraph cables. It could not have happened at a worse time. Winter weather had set in, making repair work very hazardous for cableships.

Cable and Wireless brought in one of their ships from the tropics but she was only equipped for tropical work and had no heating system. Steam heating had to be installed but even then the West Indian crew could not stand the cold weather.

There were ten cableships at one time fishing for various cables. A tremendous mud-slide had buried the cables making it impossible in some cases to locate them. One ship was fishing for thirty days for one of the cables without success. In the end she had to give up and look for another cable.

About one hundred miles of Pacific Cable Board cable was missing. A new section of cable had to be laid and spliced in. Ships were covered in ice and crews had to be set to work chipping it away.

On another occasion in the 1950s, Canadian Teleglobe's cableship *John Cabot* was faced with an emergency situation when one of its mini-subs got into difficulties at the bottom of the ocean. It was touch and go for a while for the two man crew as the air supply neared exhaustion. The *John Cabot*'s other mini-sub went down and rigged wire cables around the sub enabling it to be hauled to the surface—just in time.

CHAPTER
EIGHT

# Wartime on Fanning Island

THE ADVENT OF WORLD WAR I PROVED the value of the Pacific cable system. A few months after the outbreak of war in 1914, a powerful squadron of heavy and light cruisers of the German navy was dispatched to the waters of the western Pacific on a marauding expedition. Its objective was to inflict as much damage as possible to Allied shipping and any other important installations.

Fanning Island, a small coral atoll in mid-Pacific, just forty miles from the equator and twelve hundred miles from Hawaii, was an important link in the trans-Pacific cable system. Approximately nine miles long and four wide, it had a large lagoon in the centre. Land surrounding the lagoon averaged a quarter of a mile wide, the highest point being only eleven feet, making it invisible to approaching vessels until within a few miles. The lagoon is criss-crossed by coral reefs with a navigable entrance on the western, or lee, side of the island. Just inside the entrance was English Harbour, headquarters of the the Lever Brothers' copra plantation, where the manager's bungalow and bachelor staff quarters were located. Scattered throughout the island were several camps or small villages for native employees and their families. The natives were not indigenous but were enlisted in the Gilbert and Ellice Islands to the west and brought to Fanning Island, in those days, by the copra schooner *Doris Crane*, as indentured labour for a term of three years.

The Cable Station consisted of a group of buildings surrounding a central compound. The top floor of a large two-storied concrete building on the western side contained the operating room, billiard room, library, manager's office and a dance hall with a player piano and gramophone. On the ground floor were the bachelor's messroom and various storage rooms.

The eastern side of the compound was lined with two long narrow barrack-like wooden buildings for bachelors. They were of single room width with glass doors on both sides so that the trade winds could blow right through the rooms. There were ten or twelve rooms to each building.

At the southern end of the compound was the doctor's residence and, on the opposite end, concrete houses for members of the married staff. More married houses were in the background. In the centre was a concrete tennis court. Coconut palms and pawpaw trees lined the compound.

Originally Fanning was served by the schooner *Tangaroa*, later the cargo ships *Wairuna* and *Hauraki*, of the Union Steamship Company in Sydney, Australia, called en route to Vancouver or San Francisco about every two or three months bringing mail, freight and passengers. They anchored off the reef at Whaler's Anchorage, on the lee side of the island near the Cable Station. Freight and passengers were discharged into lighters then rowed ashore through the surf to a landing on the beach.

When a ship was expected, a native was sent atop the one hundred foot flagpole to keep watch. When he sighted the ship he called out "Sailho!" The cry was taken up by all who heard it and repeated far and wide. Everyone who could, gathered at the landing place to welcome any new members of the staff, or just to watch the activity on this red-letter day.

One day in September 1914, Alfred Smith, superintendent of the Cable Station, was seated in his office in the concrete building which housed not only the operations room but also various other departments of the complex. Most of the time his duties were not onerous, but his responsibilities were considerable. Besides technical matters, they included the health and welfare of about thirty employees, six of whom were married men with families of small children.

Informed by head office in London that a German Naval squadron was cruising the waters of the western Pacific, he was advised to take precautionary measures in case the Cable Station was one of its objectives.

What kind of measures could he take? The island was defenceless and the nearest civilization at Hawaii was over a thousand miles away. The only thing he could do was to bury spare instruments and a cache of small arms and ammunition at various places nearby. Having done that, he drew a plan showing where the articles had been buried and put it in the office safe, which also contained several thousand gold sovereigns used for paying the staff each month.

He was kept advised of the movements of the German cruisers by London office, who received the information from the Royal Navy. At the end of August he received a message stating that two ships, the *Nuremburg* (a.k.a. *Nurnberg*) and the *Leipzig*, were apparently bound for Fanning Island and were due there any day.

Smith gave orders that a native with binoculars be stationed atop the one hundred foot flagpole during daylight hours to keep a lookout for ships.

A few days later there was the cry of "Sailho" from the lookout. "I wonder what ship it is?" Smith asked his male secretary, adding "See what flag the ship is flying."

The native reported that it was a warship flying the French flag. Much relieved, Smith, who was also the local District Commissioner, put on his pith helmet with its brass knob, symbol of his authority and, after the ship anchored off the reef, went down to the beach to welcome the boat coming ashore from the vessel. Members of his staff and their families flocked to the beach, for the arrival of any boat at that isolated island was an important event.

Passengers making a landing in this way were usually carried ashore by natives so that they would not get their feet wet, since the heavy surf that battered the reef did not permit the construction of any kind of wharf. This time however, the officer in charge of the boat jumped ashore from the bow of the boat before it landed and pointed a revolver at Smith's stomach saying, "You are my prisoner!" At the same time the French flag was pulled down from the warship's mast and the German flag was run up. The officer politely regretted the necessity of restraining the staff, explaining that his orders were to destroy the cable station and that the staff would not be harmed if they did not interfere.

"Well, you needn't poke that revolver in my stomach," said Smith.

From there on relations were fairly amicable since there was nothing the staff could do to save the station. A detachment of marines armed with machine guns surrounded the station. A demolition crew blew up the powerhouse containing generators and accumulator batteries that supplied direct current to the station. Then they went through the operating room and destroyed the delicate instruments with axes. In the battery room below the operating room they began smashing Leclenche cells that provided the sending battery for the cables. The sal-ammoniac liquid spilled on the floor, and not wishing to wade in this corrosive liquid, they abandoned the task. Seeing a door on the other side of the room, one sailor asked "What's that?" A cable man replied, "Oh, that's just a meat safe." The German looked at the mess on the floor and decided he was not going to get his feet wet in that and left. Inside that 'meat safe' was the artificial line which had to be kept at a constant temperature. It was composed of capacitance and resistance that equalled that of the cable itself and enabled the cable to be worked duplex—that is in both directions at the same time. Had it been destroyed it would have taken months, possibly years, before it could have been replaced and re-balanced.

While the destruction of the station was in progress, Hughie Gregg, a native overseer on the copra plantation, chanced to return from a duck-hunting expedition. Smelling coal smoke and hearing two explosions, he hurried to the cable station. He saw the *Nuremburg* lying close inshore and two of the ship's boats pulled up on the beach. Having destroyed the essential parts of the station, the Germans were engaged in cutting the Fanning-Suva cable. The whole of the station staff and their families were watching the operation—some taking photographs of the historic event.

The *Leipzig*, which was disguised as a collier, had already cut the Fanning-Bamfield cable on the windward side of the island and dragged it out to sea in deep water. She was now engaged in doing the same to the Fanning-Suva cable. However, this time the cable was not dragged out far enough to prevent recovery. The coral reef surrounding the island was a few hundred yards wide. At high tide it was covered by the sea to a depth of about thirty feet. Beyond this the edge of the reef dropped off precipitously to unknown depths. Had the Germans pulled the cable over the edge of the reef, it could only have been raised by a specially equipped cableship.

Before leaving, the Germans opened the office safe and found the plan showing where the spare instruments and ammunition had been buried. They destroyed them and confiscated the gold sovereigns.

After the Germans departed, an assessment of the damage was made. It was decided to try to raise the cable so that a temporary connection could be made with Suva. Hughie Gregg, who was a practical man, fashioned a grappling hook out of a pickaxe and went out on the reef in a small boat to grapple for the cable. After several attempts he succeeded and raised the cable to the surface and tied a rope around it. It was then hoisted on to a raft and a wire connected between the core of the cable and the cable office. A temporary contact was made with Suva and an explanation given of what had happened, together with a list of equipment and apparatus required to put the station in operation again.

The cableship *Iris*, stationed in Auckland, sailed for Fanning Island as soon as she could be refitted, to repair the cable and reconstruct the station. En route to Fanning Island she called at Suva, Fiji, where engineers had been working day and night to prepare apparatus to replace those damaged by the Germans. After replenishing her supplies she picked up this apparatus and sailed for Fanning. It took several weeks before normal operations could be resumed. In the meantime the cable staff had nothing to do but enjoy themselves, playing tennis, sailing outrigger canoes in the lagoon, swimming, etc.

In retrospect it would have been a simple matter for the *Nuremburg* to have stood off the island and demolished the cable station by shellfire. Instead the Captain chose to destroy the station with a minimum amount of danger to the staff.

After leaving Fanning Island, the German squadron headed for the Falkland Islands, at the southern tip of South America, the intention being to establish a naval base there. On the way it encountered a small British naval squadron near Cape Coronal consisting of the *Good Hope*, *Monarch* and the fast light cruiser, *Glasgow*, which was en route to Esquimalt, Victoria, B.C. to augment services there. The *Good Hope* and *Monarch* were both sunk by superior German firepower, while the *Glasgow* escaped after a running gun battle and informed the British Admiralty of the disaster. The Admiralty then dispatched a strong naval force to the Falkland Islands to deal with the German squadron. In a famous naval battle, the German squadron, including the *Nuremburg*, was completely destroyed.

CHAPTER
NINE

# Wartime Bamfield

AT THE OTHER END OF THE Fanning Island cable was the small, isolated fishing village of Bamfield. It was located on the south-eastern shore of Barkley Sound, thirty miles by water from the logging town of Alberni, on the rugged west coast of Vancouver Island. Enclosed by the primeval forest on the east, and the waters of the Sound on the west, it was accessible only by water. The only means of public transportation to the outside world was by mail-boat to Alberni three times a week, and by small coastal steamers three times a month.

Prior to 1900, Bamfield was a small native village with only a few white pioneer settlers. After the Cable Station was built in 1902, it became the most strategic place on the west coast of Vancouver Island. The main building on the station was perched high on a cliff overlooking Barkley Sound. It housed not only the instrument and maintenance rooms but also accommodated about twenty bachelors and a dozen Chinese servants. There were about fifty rooms altogether, including a large dining room with picture-windows commanding panoramic views of the Sound. In addition there were: a library containing over three thousand books, a billiard room and a music room. On the spacious grounds of the station compound were two concrete tennis courts and a dozen houses for the married staff.

When war was declared on 4th August, 1914, a company of Royal Engineers was immediately dispatched from Victoria, the capital of British Columbia, to guard the Cable Station from sabotage. They were temporarily accommodated in the basement of the station until tents arrived. The station was encircled with barbed wire with an entrance guarded by sentries. Next to arrive—unexpectedly—were twelve civilian censors who were somehow squeezed into the staff bachelor quarters. Their job was to decode all cable messages, seeking information that would be of value to the enemy.

When the Royal Engineers finished their work of making the station secure, they were replaced by members of the 88th Victorian Fusiliers as sentries. No one could approach the station without being brought to an abrupt halt by the command at rifle point. Only those who knew the password were allowed to enter.

Meanwhile the cable staff, who could not be reinforced because of their special qualifications, worked twelve hours a day, night and day, seven days a week to cope with the enormous increase of messages due to the war. When not on duty, their lives consisted mainly of eating and sleeping. The constant strain, especially for those on all-night duty who had difficulty sleeping in the daytime, was too much for one man who had a nervous breakdown and had to be transferred elsewhere.

The work of the operating staff consisted of relaying messages from Australia and New Zealand to North America and Britain. At this stage of development all messages had to be relayed by hand since, as yet, no automatic relays had been devised to translate signals from the cable to the landline across Canada—two entirely different systems. The operators were human relays re-transmitting wavy signals from the cable to dot and dash holes in a waxed transmitting tape, using small rubber-tipped mallets, one in each hand, to depress tappets on a punch block. This tape was then fed into an automatic transmitter which sent the signals over the long cross-Canada landline to Halifax, then by cable to London.

The speed of the cable at that time was only 115 letters per minute, so the traffic backed up in huge quantities at the terminals. Cheap rate messages took weeks to reach their destination since priority messages had preference and were expedited. Sometimes, under favourable conditions, the speed would be increased to 125 letters per minute—especially when there was a long spell of plain language messages, such as casualty lists or pay records. These messages were very long. There was one of 19,000 words which was divided into sheets of 50 words each. It took about a week for that one to be relayed. Being a cheap rate message it took second place to ordinary full rate messages. Most full rate business messages were in ten-letter code words and had to be de-coded by the censors.

Received signals were gummed down on a pad and passed over to the transmitting operator. At this time, the method of transmission was in a period of transition. A change was being made from hand stick perforators to Kleinschmidt perforators which had a keyboard like a typewriter. Some operators, using the old-fashioned stick perforators, found it difficult at times to keep up with the fixed speed of the transmitter, so they were gradually training themselves to operate the faster, easier Kleinschmidt perforator on these long plain English casualty lists. This meant learning to touch-type, since one could not take his eyes off the signals to look at the keys, but had to keep them fixed on the signals.

When the Germans cut the cables at Fanning Island, the whole service between Australia, North America and Britain, via the Pacific, was interrupted. It was a welcome relief for the Bamfield staff who had nothing to do until restoration. They hiked over the forest trails, went hunting and fishing and enjoyed a well-earned holiday.

The Cable Station, perched high on a cliff overlooking Barkley Sound, provided a good observation post. One day the staff saw two cruisers anchored in the Admiralty anchorage behind Wizard Reef, about five miles from the station. These were the *Idzumo*, of the Japanese Navy and the British light cruiser *Newcastle*. They appeared to be on some kind of patrol duty since they came and went frequently.

The ships excited the curiosity of the staff, and one morning, Angus Fraser, chief electrician on the station, suggested to a group at breakfast in the messroom that they go in his launch to have a closer look at the warships. As it was a dull, cold morning with a choppy sea, they dressed in mackinaws and gumboots and climbed into Fraser's open boat. As they approached the *Idzumo*, they were preparing to circle the ship when they noticed a ladder being lowered with a petty officer standing by. He beckoned them to come alongside, which they did somewhat apprehensively, fearing a reprimand. Instead they were received with bows and were requested to come aboard, which they did. They were received on deck by the officer of the watch, again with bows, and conducted below to a spacious cabin where the bearded captain politely bid them, in perfect English, to be seated.

He opened conversation by lauding the Allied efforts against the enemy and said he hoped for a speedy victory. At the same time he regretted that there appeared to be a feeling of mistrust, especially among Australians, of Japanese motives in the Pacific.

Were any of his guests Australians, he asked politely. Two shabbily garbed members shifted nervously and admitted they were Australians. The captain thereupon assured them, on behalf of the Imperial government and the Emperor himself, of the affection and admiration with which the Australians were regarded and that the very thought of hostility shocked the Japanese people.

The arrival of the steward with sandwiches, drinks of all kinds, and cigars and cigarettes, which were handed around with profound bows, changed the subject to lighter matters. After many rounds of drinks the party became quite hilarious. Before it got out of hand, Fraser rose to make a polite speech of thanks and suggested that it was time to leave.

The Captain seemed reluctant to see them go and handed them over to a junior officer who conducted them on a tour of the ship. They were shown the spotless kitchens, the crew's quarters, gun crews asleep in their blankets beside the guns.

They departed amid bows and expressions of goodwill in a very warm manner.

The group then approached the *Newcastle* expecting a similar welcome, but they were disappointed. The sailors on board gave no sign of recognition, so they returned to the station and related to an envious audience in the library, the tale of their day's adventure.

After the battle of Cape Coronal, where the *Good Hope* and *Monarch* were sunk by the German squadron, the Bamfield Cable Station was deluged with urgent coded messages for the *Newcastle*, still anchored in Barkley Sound. The messages were handed to two naval signallers who relayed them to the *Newcastle* by means of blinking Morse lamps located on the station verandah. Next morning the anchorage near Wizard Reef was empty, both vessels having departed for unknown destinations.

After the cables at Fanning Island were repaired and communications with Australia restored, the staff at Bamfield returned to war-time conditions, working twelve hours a day for the duration of the war.

In that exhausted condition they were totally unprepared for the devastation that followed.

CHAPTER
TEN

# Influenza Epidemic

WHEN WORLD WAR I ENDED IN 1918, disaster struck Bamfield. The world-wide epidemic of influenza wrought havoc with the cable staff, most of whom were stricken with the disease. As no medical aid was available in that isolated community, the sick were attended to by the stationary engineer, George Oliver and his wife. After finishing the nightly charging of the station's accumulator batteries and shutting down the engine at 11 p.m., he and his wife visited every room in the Bachelor Quarters to attend to the sick—often staying until the early hours of the morning with serious cases.

The climax came when only five operators out of a staff of twenty-five were well enough to work with the manager, Robert Bain, to cover the whole twenty-four hour period. Conditions were so serious that a doctor and two orderlies were sent by the army in Victoria, and five cable operators were transferred to Bamfield from Halifax to relieve the situation.

The doctor decided to move two married men from their houses to rooms in the basement of the Bachelor quarters—possibly to prevent their families from catching the disease. Subsequently it was learned that it was fatal to move any patient. It proved to be so in these cases.

Another man who died was a young New Zealander, Don Hay. He shared ownership in the launch *Anzac* with another operator named Fred Andrews. The Patterson girls from Alberni were visiting Bamfield at the time and wanted to return so Andrews and Hay decided to take them back in their launch. By the time they reached Alberni six hours later, Hay was running a high temperature and died a few days later.

One of the relief operators sent from Halifax was an English teenager named William Fowlie. Having recently graduated from the London

Training School for cable operators, he had been transferred to the Pacific Cable Board's station in Halifax, Nova Scotia, in February 1917, as a full fledged cable operator. Still in his late teens, he had been excited at the prospect of serving overseas.

Only ten months later, he was badly injured in the devastating Halifax explosion when, during the war, a fully laden ammunition ship caught fire and exploded after colliding with another vessel in Halifax Harbour. The gigantic explosion almost erased the city of Halifax. Fowlie had just come off night duty and was having breakfast in his boarding house when the explosion occurred. He was flung from his chair and a heavy door fell on top of him. Clad only in a business suit, he lay injured and exposed to the bitterly cold winter weather for many hours before being rescued. Suffering from cracked ribs as well as pleurisy and pneumonia, he lay at death's door for many days before gradually recovering.

The following year he was involved in another crisis: the influenza epidemic. It was so bad in Halifax that all community gatherings were prohibited. It was the same all across Canada—even to the remote fishing village of Bamfield. Fowlie was one of the five cable operators sent there to help relieve the situation. He arrived on a dull day in November, 1918.

His account of his arrival, given in an interview many years later, follows:

> I arrived in the evening and went straight to the Bachelor Quarters, which at that time also included the operations room. Some of the staff came out to greet us. Everybody was depressed. Three members had died of the 'flu. The epidemic was so bad in Alberni that Dr. Hilton had been unable to come down to Bamfield so Miss Wilson, a nurses' aide, who happened to be in Bamfield, was looking after the staff.
>
> Most of the staff members were recovering from influenza and not really fit enough to be on duty. As a special consideration of the circumstances, authorities in London head office had ruled that, if not well enough to work eight hours, perhaps they could work four, and that is what some of them were doing. They should not have been working at all but as cable operators were irreplaceable there was no alternative.
>
> Operators were allowed to make only one error in seven thousand words while transmitting and receiving messages. If that percentage could not be maintained for a year, a man lost his annual increment and was fined. Well—the Pacific Cable Board generously stated that if a man could prove

that he lost his increment while working during the influenza epidemic, those months could be stricken from the record and would not count against him.

In addition to the operating staff there were a dozen censors. Each coded message passing through Bamfield had to be copied and then decoded. The translation was kept on file so that a list of persons trading with the enemy could be compiled from this information.

I arrived at Bamfield late at night. I reported for duty next morning at ten o'clock. I still remember my first impressions. Mr. Phippard, who was supervisor on that shift, was very smart and alert. He wore a navy blue suit, and, if I remember rightly, a stiff collar. However the general effect was rather spoiled by wearing a pair of soft slippers. Robert Bain was the manager. I had previously worked with him in Halifax and knew him fairly well. Mr. Price was clerk-in-charge, Danvers Osborne was next and Phippard was the junior.

I managed to get through that first day. My operating was put to a severe test by the difficult signals but somehow I struggled through the day. I had dinner in the Bachelors' Mess that evening and went for a walk on the Lifesaving Trail afterwards with Danvers Osborne. He was rather an eccentric Englishman with connections to the aristocracy. He used to write letters to the paper about the Lifesaving Trail on the west coast.

The staff gradually recovered from the effects of the 'flu and, during that first summer, we played tennis, went salmon fishing, hiked on forest trails and enjoyed the freedom of a normal life.

I should mention the tragic death of one of the censors named Harrop. He had an attack of appendicitis in the middle of the night. His wife came over to the office to see if someone could take him to Alberni, thirty miles away by water. Several of the men owned boats but, of course, they couldn't do anything until they came off duty. So she rowed across the darkened inlet to the lifeboat station. Unfortunately the tide was so low that the lifeboat could not be launched — the slipway was too short. She returned to the station and waited until the men came off duty at 10 a.m. Then they had to have breakfast and get their boat ready. He was taken up to Alberni in the *Kiwi*, but it took six hours to get there and by that time his appendix had burst and it was too late.

His widow later became my wife.

Asked to name some of the staff, he said:

While I was in Bamfield the staff consisted of Angus Fraser, who was Chief Engineer and lived in the house on top of the hill. Later the McIntyres lived there. The Moores occupied the house close to the trail by the tennis court.

We liked having Bill Whaley on the staff—he was always good fun. Eric Beehan and McCombie were transferred to Fanning Island. Jock Baird came back from the war and Mrs. Baird came up from Fiji to join him. Jerry Kerr was here for a while but he resigned and joined the Commercial Pacific Cable Company. Sammy Williams was very artistic. He and Guy Neville organized amateur theatricals. During one of these someone composed the following ditty, to the tune of "What a friend we have in Jesus":

> We don't want your Fanning Island,
> We don't want Bamfield any more;
> Take me back to sunny New South Wales,
> I never missed me 'ome so much before.
> We don't want your Nova Scotia,
> Or the islands of Fiji;
> Take me back to sunny New South Wales,
> Sydney is the place for me.

Fred Harry was engaged to Miss McKay who died shortly after Fred was transferred to Fanning Island. Roy Weston, who married a girl from Alberni, was a great mechanic. Percy Cuthbert was here. His father was the first manager in Bamfield.

Then there were Reg and Frank McCay, Croger, Mann—a great man for the ducks—Ed Meade, a Newfoundlander who came to us from the Eastern Cable Company, Chuck McCay from Auckland, Patterson—a nephew of the Pattersons of Cape Beale lighthouse—the two Andrews brothers, Alf Smith who came here as manager, Trot Smith, who had been in the ranks of the Australian Armed Forces. S. G. Hill and Jock Baird had been commissioned officers. Alf Smith, the manager, was rather annoyed because these three men wore their army uniforms in the office and Smith found it rather difficult to enforce discipline when they were wearing officer's uniforms. There was another Smith nicknamed Trot who cut cordwood in his spare time and sold it to the station for seven dollars a cord. Then there was Cogan who could play the piano very well. He was also good on the Kleinschmidt perforator, applying the same technique of playing the piano to the Kleinschmidt. Some operators could play popular tunes on both the typewriter and the Kleinschmidt while operating.

Asked about operating conditions, Fowlie said:

Most commercial traffic, which consisted of about half the total traffic, was in five letter code words. A combination such as ALAVK could mean "your order of October 25th on hand," or something like that. Two combinations of five letters were permitted in one ten letter word. This provided considerable savings to business firms, since one full rate word cost about two shillings, or fifty cents.

> We were using Morse inkers on the landline side. They were not as good as the undulator recorders that replaced them in later years. The main landline across Canada to Montreal was only a physical line. The carrier, which permitted a higher speed and more channels, had not yet been introduced.
>
> At Montreal the signals were re-transmitted to Halifax and then re-transmitted to London across the Atlantic cables. Automatic through working came later.
>
> Reading and transmitting both cable and landline signals was hard work requiring a lot of concentration. I found night duty especially trying because I have always been prone to drowsiness. How I managed to get through without making too many errors, I do not know.
>
> Jack Turnbull was here with me at this time. We both got married in Victoria about the same time. He married Moragh Jameson of Victoria, and I married the widow of one of the censors. He went on leave to Australia and my wife and I went on to England.

In recognition of the supererogation displayed by the Bamfield staff during the war years, the Pacific Cable Board built a community-type hall for the benefit of the staff. It was a combined dance hall, badminton court and was equipped with a stage for amateur theatricals. A projection booth for movies was added later.

Not everyone could adapt to the isolation of remote cable stations in the early years.

The Alberni *Pioneer News* reported on 5th March 1910, that W. J. Barry, an operator at the Bamfield cable station was swept to his death by a violent gale in Barkley Sound on the west coast of Vancouver Island.

It reported that Barry, while working on night duty, had exhibited signs of mental instability. He was released from duty and advised to go to bed. However he expressed his intention of going out for a row in his rowboat. His colleagues tried to dissuade him but he ignored them and hurried down the hill to the wharf. Shoving off a boat he climbed into it. In his excitement he forgot to take a pair of oars. Helpless, he was quickly swept out of the harbour by the strong winds. Hearing his cries of distress, his colleagues launched a boat in an attempt to rescue him. However they were unable to locate him in the darkness and, after struggling with wind and waves, were compelled to return.

The lifesaving crew was notified and turned out with alacrity. J. B. McKay, the telegraph lineman, also commenced a search in his gasboat, but Barry was never seen again.

CHAPTER
ELEVEN

# Doubtless Bay Cable Station

THE ONLY SURVIVING MEMBER OF the Doubtless Bay cable staff, Robert Marchant, gave the following description of the early days on that station:

> I joined the Pacific Cable Board in 1909. The Superintendent at that time was Alf Smith. We were using a separate morse landline between Doubtless Bay and Auckland. The New Zealand government received a pro rata rate on all messages transmitted over this line.
>
> After being on clerical work for a year, I managed to pass the operating exams and qualify as an operator. I can thank the senior operators for this. I was given a month's trial of operating and, as my error percentage was satisfactory, I was put on operating. Milward, who was then Manager in the Pacific in Sydney, granted me an operator's salary, but said I would have to complete two years of service before I could sign the Pacific Cable Board's staff agreement.
>
> In those days it was mostly keying work, or using the old two punching irons to perforate the tape which was then put through the old hand-wound clockwork transmitter for onward transmission.
>
> We re-transmitted all the P.C.B. international cablegrams from Sydney to Suva and onwards. The cable from Sydney to Doubtless Bay was connected on shore to the cable going north to Suva, Norfolk Island, Fanning Island and Bamfield. In 1912 a cable was laid from Mangonui, Doubtless Bay, around the coast to Auckland and into the post office, doing away with the morse landline. During World War II, the cables were joined at sea instead of lying buried ashore at Doubtless Bay.
>
> We were really comfortable at Doubtless Bay, having good quarters with tennis court, hall, billiard room and a smoke room with a piano which was sometimes treated to a glass of beer. The Mangonui Hotel was three or four miles away, so a person going there hired a horse because the road was knee deep in mud in places. Often a riderless horse would return to the station

and, half an hour later, its rider would appear over the hill staggering along covered in mud from head to foot. The usual procedure was to dump him into the bath and then into pyjamas. If he was required on duty, someone else would do his shift for him.

I left Auckland in 1914 and went overseas with the 3rd Australian Military Regiment as a trooper. In Egypt I met Joe Nevitt, an Aussie P.C.B. operator who was in the 2nd Australian Light Horse. I returned from the war after being wounded and arrived home three days prior to peace being declared. I applied for reinstatement but the P.C.B.'s doctor could not pass me as I still had a piece of shrapnel under my spine.

I have it in black and white on a form with the P.C.B. monogram stating that I was a deserter for leaving the Company in the hour of need. It takes two years to train a cable operator, the memo said, and only three months to train a soldier.

## Note

At the start of World War I, some P.C.B. staff in New Zealand and Australia resigned and went overseas with the New Zealand and Australian signalling corps. When conscription was brought in for all fit men between the ages of twenty and forty, P.C.B. staff were classified as essential service, putting a stop to further resignations. P.C.B. staff were issued with brass badges denoting essential work.

CHAPTER TWELVE

# Sydney 1922

After completing his training in Southport each boy, upon arrival in Sydney, was placed on one year's probation and signed a five-year contract regarding his terms of employment. He then received a salary of one hundred and fifty pounds sterling per annum—the equivalent of $729 at that time. Also he was given a copy of the official handbook of working rules and regulations, in which emphasis was placed on the fact that, on and off duty, he must remember his responsibilities as a "Gentleman on Imperial Service" and govern himself accordingly.

The handbook gave the various speeds at which cables were worked and the number of words that had to be handled each hour. At a speed of two hundred letters a minute, it was 2,400 words per hour. It also stated, "Senders not generally securing the above earning power of the several listed speeds should be called upon for report, and the incident brought to the attention of the manager."

At this time, methods of transmission and reception had not changed much since their original invention. All traffic between Sydney and Melbourne was transmitted by key on a manually operated overland telegraph circuit. The three operators who manned the circuit around the clock were, Karnaghan, Fletcher and Alexander. They could carry on a conversation with a bystander after Melbourne had started sending traffic then, casually inserting a form in the typewriter, start receiving the message several lines behind—still carrying on the conversation.

Messages were received over the Auckland-Sydney cable, which was the main international route at that time, by means of a siphon recorder, the wavy seismograph-like signals being recorded on a moving tape which passed over a typewriter behind a thick glass tube filled with water which magnified the signals to a desirable size. Messages were typed up by an operator and were checked by another operator sitting beside him. In case of doubtful signals they could confer with each other, or with the supervisor. Press messages required a form with ten carbon copies for distribution to various press services. Imagine how hard one had to hit the typewriter keys in order to make ten legible copies!

Transmission on the cable was by means of a keyboard perforator and a Cox transmitter. The cable was worked at a speed of two hundred letters a minute and that speed had to be maintained by the operator regardless of what difficulties he might have reading hand-written cablegrams, which were sometimes almost illegible—especially code messages. He could not stop the transmitter without attracting the attention of the supervisor. An idle transmitter meant lost cable space that never could be regained. Of course he also had to initial and time each message and change the roll of tape in the perforator before it ran out.

Not only did the operator have to keep the cable full, he also had to be accurate. If he did not maintain the required standard of accuracy, a check was made of each succeeding month's work and his annual increment was not recommended until his record for the preceeding twelve months came within the scheduled allowance. Delayed increments entailed a permanent loss of seniority. The monthly percentage of errors for junior operators was .025 and for senior operators, .012. The system produced some of the best operators in the world. An experienced P.C.B. operator could get a job with any other cable company—at a higher salary. When the new All America Cable Company was formed in South America, attempts were made to attract P.C.B. operators to serve in that area. A few accepted the offer.

The Pacific Cable Board's sphere of operations extended from Sydney to Halifax, Nova Scotia, thence across the Atlantic to London. The Pacific route was: Sydney-Auckland-Suva-Fanning Island-Bamfield-Halifax, the latter section being an overland telegraphic line maintained by the Canadian Pacific Railway Company. There was an alternate route from Sydney-Southport-Norfolk Island-Suva, used only when the main route was interrupted. The connection between Sydney and Southport was a land-line circuit maintained by the N.S.W. Government Telegraph system.

We worked shifts around the clock. When it was night time in Sydney, it was day time in London. Sometimes, during clear periods in the early hours of the morning, before the day's traffic had begun to flow in from London, we used to go outside and chat with the policeman on the beat. The streets were deserted except for the odd straggler wending his way homeward on foot—since all public transportation had long since ceased for the night. A cup of hot tea and a few slices of buttered toast made on an electric radiator fortified us for the expected flood of incoming messages.

As there was no form of central heating in those days (and not much in Australia today) we had to wear overcoats to keep warm while operating on night duty. The only heat provided came from a few small electric radiators.

When a vacancy occurred at any station on the line—especially isolated places like Norfolk Island, Fanning Island and Bamfield, the Board preferred to send a volunteer. Only when there were no volunteers was a man told to go.

I had been on Sydney station less than two years when a vacancy occurred for a bachelor on Fanning Island. Normally only senior operators were sent to that isolated station because the signals over the long cable to Bamfield were difficult to read.

No one volunteered so I decided to do so. The next day I was called into the assistant manager's office.

"You want to go to Fanning Island, Scott?" he asked.

"Yes, Sir," I replied.

"How old are you?" he asked.

"Eighteen, Sir," I replied.

"Eighteen?" he exclaimed. "Eighteen! Hmmmm! Hmmmm!" while walking up and down as he considered the matter.

"Well," he said after a few moments, "I don't see why not!"

And so I became the youngest operator ever to go to Fanning Island.

Again there was the excitement of preparing for a transfer, this time to the most exotic station in the service—an isolated coral atoll. My widowed mother took me to a department store and bought a tropical outfit for me. Within a few days I entrained for Newcastle where the cargo ship *Wairuna* was coaling. I stayed aboard for a few days until she finished coaling, then steamed out of the harbour and over the horizon for Fanning Island, 4,000 miles away.

CHAPTER
THIRTEEN

# Fanning Island 1923-1925

MY VOYAGE ON THE "WAIRUNA" to Fanning Island was exciting. The weather was good and the blue, blue ocean sparkled in the tropic sun. Most of my time was spent in the heaving prow of the vessel watching the bow waves where porpoises cavorted and flying fish skimmed over the surface of the water for considerable distances before falling back into the sea. One flew through the porthole of my cabin and landed on the floor. The steward served it for breakfast next morning.

I became friendly with the radio operator, often visiting him in his combined radio shack and cabin. He explained that he had to provide his own radio receiver because the one supplied by the company was not very sensitive. The transmitter was the old-fashioned spark-gap type.

Aboard with me was another cable man, Harold Williams, also bound for Fanning Island.

We arrived off the island about two weeks later, and disembarking into a lighter, were rowed ashore through the surf. Brown-skinned islanders carried us ashore on their backs. (This was not a welcome; it was a service provided to all who came ashore.) We were welcomed by other members of the staff and taken to our rooms in the bachelor quarters. These consisted of two long narrow barrack-like wooden buildings. They were of single room width with glass doors on both sides. There were ten or twelve rooms to each building. Cold showers were provided in an adjacent building, but as the pipes ran close to the surface of the ground, the water was usually hot, being heated by the rays of the sun.

The cable staff consisted of about six married men with families, twenty bachelors, an engineer named McNeill to maintain the D.C.

power plant and refrigerator unit, a handyman named Chan, a messman named Whale, and doctor E. K. Edie with his wife Jean and son Mac.

Since the island was so isolated, the technical staff received an isolation bonus and were required to serve only two years before being eligible for transfer elsewhere in the service. Some chose to remain longer.

Since the cables were in continual operation, either in one direction or the other, the staff worked around the clock in six hour shifts, five men to a shift. The Fanning Island-Bamfield cable, which was over four thousand miles long, extended northwards to Canada, and the Fanning-Suva cable, over two thousand miles southwards. The long lengths of both cables made signals difficult to read.

Tropical insects were a nuisance—especially mosquitoes. A bucket of kerosene was placed outside the operations room door, and as each man entered, he dipped his bare feet and hands into it and rubbed it over his face and arms to ward off the insects. Geckos, small translucent lizards, scampered over the walls and ceiling of the operating room. Every now and then one would lose its grip on the ceiling and thud to the floor.

On the night shift, men were served breakfast cooked by the native office boy (native servants were always referred to as boys regardless of age), which they ate while on a slack circuit.

The bachelors on night shift were awakened by a man on the evening shift. They slept under mosquito nets with only a sheet to cover them. Sometimes, if one was difficult to awaken, a land crab might be slipped under the sheet to do the job. The land crabs, about eight inches wide, ran all over the station compound, pathways and beaches. They burrowed in the earth like rabbits with the result that one might suddenly sink to his knees while walking. At night they were so numerous that they were trodden underfoot. Anyone who chanced to fall asleep on the lagoon beach at night would be awakened by crabs crawling all over him.

Besides land crabs there were also giant coconut crabs about twelve inches in width. They were called coconut crabs because they climbed the coconut palm and nipped off coconuts. Descending to the ground they ripped open the nut and ate its contents. Natives used to catch them by tying a tuft of grass around the tree trunk. When the crab descended backwards, it let go when it felt the grass, thinking it had reached the ground. The natives, who liked crab meat, then pounced on it, taking care to avoid its immense claws, which were powerful enough to break the neck of a beer bottle.

Since the island was only forty miles from the equator, days and nights were of equal duration, the sun rising and setting in the same places each day. The annual mean temperature was 83.5 degrees Fahrenheit, modified by almost continuous trade winds that shifted between southeast and northeast for varying periods.

A trail, surfaced by broken sea shells, encircled the lagoon connecting various native camps. Cable bachelors usually went barefoot, dressed only in shorts, singlet and hat. Soon after his arrival an unsuspecting man would be invited to go for a walk along this trail—in bare feet. Unlike the others his feet were tender and it was not long before he was limping painfully. After a few months barefoot, the soles of his feet became calloused and he was able to walk over anything.

Off duty hours were spent in a variety of ways: exploring the swamps and sloughs, sailing outrigger canoes, fishing in the lagoon or outside on the reef, tennis, shooting ducks and curlew, which were migratory birds in season, photography—which entailed developing and printing your own films, and swimming.

A swimming pool had been carved out of the coral reef on the ocean side of the island. It was safe when the tides were lower than medium, but on the higher tides sharks were likely to enter it. I had a small shark about two feet long hit me on the heel on one occasion. It was too small to take a bite. On another occasion I saw what seemed to be a colourful snake-shaped shell. I was about to touch it when I thought, better to poke it with a stick first. It was just as well I did because it was a large clamshell that closed with a snap when touched with the stick.

There was a fair amount of social life including dances with music supplied by gramophone or player-piano, billiards, a library with magazines from London and elsewhere. Of course, there was no radio or television at that time.

For customs' purposes, the island was considered to be a ship at sea and so was duty-free. Each man kept a well-stocked cabinet of liquid refreshments—not so much for his own consumption as for the entertainment of others. At Christmas time, Lasseters, a well-known department store in Sydney, used to send several cases of Moet and Chandon champagne to the staff in appreciation of business done with the store during the previous year.

Occasionally one might imbibe a little too freely before going on night shift and thus be unable to do his shift. All he had to do was to wake up another man and ask him to do his duty for him. He was never refused—even though it meant his relief having to work two shifts in one day. The favour would be returned at some other time. Such was the esprit de corps.

## Island Wedding

A marriage on Fanning Island was a most unusual event. Although there were about twenty bachelors on the cable staff, there were never any young ladies among the married staff's families—or anywhere else on the island. However the manager of the Copra Plantation, Major Burn-Callandar, whose ramrod straight bearing indicated his military background, had become engaged to his fiancee in England. She travelled alone by ship and train to San Francisco where she embarked on the copra schooner *Doris Crane* for the three week journey to Fanning Island. The captain of the ship, a man named McCullough, was a typical hard-drinking, hard-working, blasphemous, come-hell-or-high-water type of those days.

In due course the ship arrived and moored at the anchorage just inside the entrance to the lagoon. The bride-to-be was met by the groom and a group of ladies from the cable station, who took her in hand and prepared for the wedding a few days later. Both the plantation and the cable station compounds were gaily decorated with bunting and palm fronds.

The wedding ceremony was performed by the District Commissioner, Methven, in a small hut thatched with palm fronds that constituted the Commissioner's office. Abbey, the plantation bookkeeper, acted as best man. Standing at attention outside the entrance to the hut, was a guard of honour composed of native policemen. After the ceremony the party wended its way to a house on the cable station where Ben Keys and his wife held a reception.

Refreshments flowed freely and everyone enjoyed themselves. Upon returning to the settlement, the wedding party was met by a throng of natives who formed a procession, led by the newly-weds, and proceeded to the manager's bungalow which nestled beneath a grove of coconut palms beside the iridescent waters of the lagoon. An idyllic site for a honeymoon.

## The Copra Settlement

In my time—1923—the copra plantation was a subsidiary of Lever Brothers in London. Its headquarters were situated on the right-hand side of the main entrance from the ocean to the lagoon. It consisted of buildings for the office, manager's bungalow, quarters for the bachelors, who at that time included Winspeare, Doc Harrison, Cullimore, Reynolds and Abbey. Also there was a lawn tennis court. It was the main centre for the native labour force with platforms for drying the copra. The copra was sent to San Francisco on the *Doris Crane*, and the extracted coconut oil was used for a variety of purposes.

Since the earth on the island was composed mainly of sand and granulated coral, it was not suitable for agriculture. Soil was imported from San Francisco as ballast on the copra schooner, *Doris Crane*, and dumped on a suitable place to make a garden. It produced green vegetables and plantains—a kind of banana less sweet and more starchy than bananas.

On the whole, relations between the settlement and the cable station were fairly amicable, although there were occasions when they were not the best. William Fowlie recalled that "There was a doctor on the settlement whose wife was also a doctor. For some reason or other, the manager of the settlement and the doctor and his wife, had a falling out with the cable staff. Although the doctor's wife and I went to Fanning on the same ship, and were very friendly all the time aboard ship, after arriving at Fanning, we never spoke to each other for the next two years. When I left Fanning, she was on the same ship and by the time we reached Vancouver, we were very friendly once again. It just goes to show how isolated conditions can affect people's attitudes to each other."

I must confess that I, too, had a falling out with the manager, Burn-Callandar, and his wife. It concerned an imprudent remark I had made about them, which was promptly relayed back to them. Next time we met I was confronted with the accusation—which I could not deny. "A fine way to treat us after all we have done for you," retorted Burn-Callandar, and that was that. I was only a teenager at the time and had much to learn regarding the pitfalls of social gossip.

## Sailho!

Ships of the Union Steamship Company of Sydney anchored off the reef at Whaler's Anchorage and discharged their cargo and passengers into lighters which were then rowed ashore by natives through the surf to a landing on the beach. When a ship was expected, a native was sent atop the one hundred foot flagpole to keep watch for it. When it was sighted he would call out "Sailho!" The cry was repeated by all who heard. They gathered at the landing place to welcome new members of the staff, or just to watch the activity on this red-letter day.

On one occasion after the ship had been sighted, the boy atop the flagpole reported that the ship, which was away off on the horizon, had sailed past the island, then turned back and steamed toward the anchorage. Being so low-lying (eleven feet at its highest elevation), the island is invisible twelve miles from shore. I went out on a surfboat with the Doctor and Superintendent. As we approached the ship the captain called out through a megaphone, "Why the blankety blank don't you anchor this blasted little island of yours?"

The Superintendent aptly replied, "Well, it's got a cable forward and a cable aft—isn't that enough for you?"

## Adrift at Sea

One evening, after having visited my friend the wireless operator on the *Wairuna* as she lay at Whaler's Anchorage, I jumped into a lighter loaded with cane furniture to be ferried ashore. One of the station supervisors, Cyril Phippard, was with me. After being cast loose, the native oarsmen could not use all the oars because the boat was overloaded with cumbersome furniture. While they struggled to re-arrange the cargo, we drifted out to sea on the strong current. Darkness had fallen and we had no navigation lights. Afraid lest we would not be seen, Phippard was about to order the natives to jettison some of the cargo when the copra settlement's launch, *Pearl*, came looking for us. Her skipper spotted us in his searchlight and towed us ashore.

Next morning the Superintendent, James Letters, asked Phippard for a report, but if he had any doubts about the incident, I was able to corroborate Phippard's explanation.

## Rotten Meat

The *Wairuna* and *Hauraki* brought fresh meat as well as other foodstuffs and cargo. On one occasion the ship had refrigerator trouble and all the meat destined for the Cable Station was found, on arrival, to be unfit for human consumption. It was all thrown overboard attracting scores of voracious sharks that thrashed the waters to foam as they fought over the carcasses. As a result the staff had to live mainly on fish and rice until the next ship arrived.

## Indentured Labour

There were several hundred natives employed by the copra company to tend the plantation, gather the coconuts, extract the white meat and set it out to dry on large wooden platforms. It was an important economic harvest.

The natives were recruited in the Gilbert and Ellice Islands for a three year term. (Fanning Island was included in 1916 in that colony, which lay far to the west.) Every three years the plantation schooner, *Doris Crane*, made the journey to return part of the work force and recruit a fresh batch. Before boarding the schooner for the return journey, they were assembled in a large circle at the settlement and addressed by the manager through an interpreter. Then, as they boarded the vessel they lined up, one by one, to say goodbye to those remaining. Once aboard they crowded the decks, some climbing in the rigging, as the *Doris Crane* cast off and proceeded under power out the passage to the open Pacific, where sails were raised and a course set for the Gilbert and Ellice Islands, which are now part of a new country called Kiribati.

On the return journey with a fresh batch of natives, the procedure was the same in reverse, the recruits being addressed by the manager and advised of the do's and don'ts.

The natives were accommodated in dormitories at the settlement and in thatched huts at various villages around the island. They had been taught to sing in harmony by missionaries in their homeland, and the sound of their melodious voices was often wafted on the still night air as they sang in harmony.

The Gilbertese natives were friendly and cheerful. The one who serviced my room, a teenager about my own age, went out of his way to please me. Each day he climbed a coconut tree and picked a coconut to place in my ice chest. Fresh coconut juice made a delicious cool drink. We could converse only in pidgin English, a mixture of English and Gilbertese languages. After I left Fanning Island in 1924, he wrote the following letter to me in Australia:

Napari,<br>Fanning Island<br>14th April 1925

My Dear Sir,

Allow me to introduce to you my friend Scott who visits your house what you stop before. I am very sorry for you pecause I remember you all time.

Yours truly<br>Ioane B.

P.S. Ko na mauri Tikote. Tiakapo.<br>(How are you Scott? Goodbye.)

## Empire Day

Empire Day was celebrated in the far-flung outposts of the British Empire in traditional manner. On Fanning Island it was Sports Day. Races were held on both land and lagoon waters. Natives, both men and women, entered into the spirit of the occasion with great enthusiasm and contested in the various races such as, egg-and-spoon, three-legged, wheelbarrow (one man holding the legs of another who raced on his hands). Then there were high jumps, long jumps, and pillow fights on a horizontal pole.

On the lagoon, members of the staff who owned catamarans (single hulled vessels with a triangular sail and an outrigger) competed with each other. They were tricky vessels to handle—especially when sailed by only one man. In order to tack, or change course, the foot of the mast had to be grasped and carried bodily from one end of the craft to the other. Woe betide the man who did not first ensure that the outrigger was on the windward side—not the lee—otherwise the sail would collapse and the laborious task of raising a wet sail would ensue. The outrigger, or pontoon, was used to stabilize the craft. Under favourable conditions it would be kept just skimming the surface of the water. If the crew consisted of two or more, one sat on the outrigger in a strong wind to prevent it from rising too high and capsizing the vessel.

On one occasion, while learning to sail my own outrigger on the lagoon, I inadvertently dropped sail and could not raise it again. I had no recourse but to get out and push the craft back to dry land over the coral reefs, a dangerous undertaking. Since there were no antibiotics in those days, any scratch from the poisonous coral could become infected and difficult to heal. One man was known to have infected sores for several years.

## Naval Visitations

Once a year the British naval base in Bermuda dispatched a warship to pay a courtesy call to outposts of the Empire in the Pacific Ocean. The main purpose of the visit was to display the flag, thus emphasizing legal possession of the land.

In 1924, HMS *Curlew* paid such a visit to Fanning Island and the officers were entertained by the staff. A dance was held in their honour, although there were only about half a dozen married women for partners.

The officers were rowed ashore by a crew of sailors who stood by the boat while the officers enjoyed themselves. Some members of the cable staff, sympathizing with the sailors, presented them with a 750 ml bottle of forty percent overproof whiskey with predictable results. It ended up with the sailors teaching the cablemen a popular ditty known as the Curlew Drinking Song:

When we go out the people shout
They're bothered if they'll be bothered about,
And they won't be bothered about.
They won't be bothered about!

When we go out we always shout
We're bothered if we'll be bothered about,
And we *won't* be bothered about!
We *won't* be bothered about!

But we always b-o-t-h-e-r a-b-o-u-t!

(Of course the navy used the word buggered instead of bothered.)

Another drinking song on Fanning was: to the tune of "Ach, Du Lieber Augustine"

> The more we are together,
> Together, together,
> The more we are together
> The happier we'll be.
>
> For your friends are my friends,
> And my friends are your friends,
> And the more we are together,
> The happier we'll be.

## Island Birds

Indigenous birds on the island were the gannet, or booby, frigate bird, fairy tern and a bird of paradise with two long trailing tail feathers. The gannets were so wild that they were tame. Seldom having seen man, they had not learned to fear him and could be picked up while sitting on their nests on the ground. Fairy terns were small white birds with large black eyes and black bills. When approached they would flutter about like butterflies.

Migratory birds like the russet-rumped curlews, and ducks, flew all the way from Siberia and other far northern places across the Pacific Ocean. Some landed on Fanning Island in an exhausted condition; they just flopped down anywhere. They were left to fatten up for a few weeks then hunted and shot as a welcome change from a diet of canned meat and fish and rice.

The frigate bird was best described by an early trader, Captain Lucatt, who visited Fanning Island in 1848. "A more appropriate name," he said, "would be a piratical hawk. It was a dastardly glutton, attacking the common gannet in mid-air while it was flying to its nest with a fish in its beak, forcing it to disgorge its prey. As the fish fell, the frigate bird swooped down and caught it before it reached the ground, carrying it off triumphantly despite shrieks of protest from the gannet."

There were no land animals on the island.

## Last Days on Fanning

As mentioned previously, it was customary for each member of the cable staff to keep a well-stocked cabinet of liquor. It consisted of whiskey, various liqueurs, beer, gin, tonic, bitters, wine, sherry, etc. For customs' purposes, Fanning was treated as a ship at sea so all imported goods were duty free. A 750 ml bottle of forty percent overproof whiskey could be bought at the staff canteen for half a crown (two shillings and sixpence) the equivalent of about fifty cents. It was not just for personal consumption, but dispensed mainly for friendly get-togethers and entertainment purposes.

On one such occasion I was entertaining a small group of senior men in my room. As usual the conversation was mainly about sporting events in Australia, cricket in particular, for it was the days of Don Bradman and Larwood and bodyline bowling.

I was sitting on the edge of my writing desk, since all chairs were occupied. Suddenly I keeled over and fell flat on the floor. I don't remember what happened after that and my memory is hazy about the next few days. I remember the staff Doctor Edie asking questions about my family history. I told him my younger sister had epilepsy and that my Father had died of a combination of "shell shock" and erysipelas after returning from World War I. From that he deduced that I had petit mal and recommended that I be transferred back home, a few months before the normal term of two years.

I left Fanning on the next steamer, MV *Hauraki*, en route for San Francisco. After a few days in San Francisco, I boarded the SS *Niagara* bound for Sydney. Although I have a vivid memory of events before this incident, I have no recollections of anything afterwards—even the voyage on the *Hauraki*. My mind was dazed.

On board the *Niagara* I became acquainted with Walter Burley-Griffin, the American architect who designed the city of Canberra, capital of Australia. I later visited the family at their home, "Castlecrag," in what were then the wilds of North Sydney.

Also on board was the New Zealand "All Black" rugby football team returning from a victorious tour of the United Kingdom.

Upon arrival in Sydney I stayed with my mother and younger sisters Patricia and Janet. My older brother Harold and sister Thelma had left home and married.

I

2

1 Southport Class of 1920. FRONT ROW: Lucas, Henkel, Swinney, Sanders, Schmidt, Vormister and Williams. STANDING: Arthur Denison, Nathan, Stan Denison, Berry and Scott.

2 Anzac Day celebrations, Southport. Included in this photo are the pier, rotunda and baths.

3

4

3 Bachelor quarters. FROM LEFT TO RIGHT: Nathan, Greenlees, Lucas, Henkel, Sparks, Schmidt and Kelso.

4 A typical trainee's room, Southport Training School.

5

6

7

5 Southport movie theatre on Nerang Street, *c.* 1921.
6 Superintendent Warner and family, 1921.
7 Visiting VIPs. The Lieutenant Governor.

8 

9 

8 Fanning Island Staff—cable and plantation, 1923. FRONT ROW: Weston, Brady, Moginie, Kennys, Methven, Burn-Callandar, Keys, Letters and Cotters. BACK ROW: Williams, Croger, Blackley, McNeil, Harrison, McCay, Collins, Reynolds, Forsythe, Scott, Abbey, Keys, Wilkinsoon, Duggan, Whale, Yeomans, Jaffray, Stubbs and Comber.

9 View of the cable station site from atop flagpole, 1923.

10

11

10 Oblique palms near lagoon.

11 Fanning Island wedding—newlyweds at afternoon tea!

12 

 13

14 

12 Author at age eighteen.

13 A beachhead of landcrabs.

14 Superintendent Blackley, Doctor Kenny and District Commissioner Methven returning from HMS *Curlew.*

15

16

17

18

15 Empire Day sports—ladies sprinting to finish line.

16 Teams of men on same day, participating in a "wheelbarrow" race.

17 Departing workers boarding the *Doris Crane.*

18 Captain McCullough and the first mate of the *Doris Crane.*

19

20

21

19 Target practise: messers. Yeomans, Collins, Weston, Blackley and Williams.
20 Outrigger sailing: Morris, Blackley and Williams.
21 Tennis tournament—sometimes mixed doubles.

CHAPTER
FOURTEEN

# Fowlie Transferred to Engineering Grade

AFTER SURVIVING THE INFLUENZA epidemic, William Fowlie and his bride left Bamfield in 1920. They spent a brief honeymoon in Butchart Gardens, in Victoria, then went on to England where he spent his leave with his bride, exploring the countryside in a motorcycle and sidecar.

After leave he was transferred from the operating grade to engineering. He attended training school at Hampstead, London, run by Via Eastern Cable Company, who had taken over a large building (formerly a girls' school) and converted it to their own use.

"I was sent there," he told me many years later, "to become acquainted with their methods and their staff. It was a most enjoyable experience. The course was excellent and the instructors were very good. At the end of three months I was qualified to go abroad, but I asked for another three months so that I could complete further studies.

"All told, I had a year in England, then I was sent back to Halifax as an assistant engineer, under a Western Telegraph engineer who was installing a new system of relay working known as the regenerator system. It was very good. It was well designed and had proved itself."

In 1925 Fowlie was transferred temporarily to Bamfield for six months. At that time a site was being prepared for a new concrete office building to house new equipment when the service was duplicated in 1926. There was a difference of opinion as to where the building should be located. K. C. Cox, the station engineer, wanted it up on top of the hill but Sherlock, the manager, and Ed Turner, the assistant manager, thought that Mawson Rattenbury, the architect who had designed the original building, should be consulted. Rattenbury thought it would look better if a shelf were blasted out of the cliff in front of the old wooden building and the three-storied concrete structure erected there. His opinion prevailed.

While these preparations for the new cable were in progress, the old Bamfield-Victoria cable broke. A section of the old German Atlantic cable which had been cut and commandeered by the Allies during World War I, it was a patchwork of various bits and pieces and broke fairly frequently. Although it was only an emergency route and not used commercially by the Pacific Cable Board, nevertheless the Canadian government insisted that it be maintained in case of need.

Fowlie tells of his experience on that occasion:

> I was sent to Victoria to contact the captain of the cableship *Restorer*, stationed in Victoria. Each of our tests in Bamfield had given different readings—placing the fault anywhere between the station and the Perouse Banks, off the Strait of Juan de Fuca. So it was decided to test the cable from both ends. I was instructed to go to sea on the *Restorer* as the Cable Board's representative.
>
> This was my first important mission as an engineer and I am afraid I got a swollen head. Since I had been transferred to Bamfield on a temporary basis, my wife and children had remained in Victoria. I was very glad to have an opportunity to see them. The following morning I walked down to the gangway of the *Restorer* and asked the sailor in attendance, "Can I see the Captain, please?" The sailor looked a bit nonplussed and went and got the quartermaster. I repeated my question and he, too, looked nonplussed and went away and came back with the Chief Officer. I am afraid I wasn't very tactful. I repeated my question in quite a firm voice, "Can I see the Captain, please?" He replied, "Don't you mean, will the Captain see you?" Well, that pinned my ears back.
>
> Anyway he and I became good friends and I went to sea on the *Restorer*. Before sailing I did a week's testing on that end of the cable to try and locate the fault, but without success. Finally we decided that it was somewhere off Cape Beale, five miles from the Bamfield Cable Station. Then we put to sea. I'm afraid I wasn't in any condition to take much interest in cable repairs. I was too seasick.
>
> The actual repairs were in the hands of Captain Spark, I think it was, on the *Restorer*. The electrician was a Mr. Porter, and the Chief Engineer was Mr. Ryall.
>
> When we arrived off Cape Beale, we cut into the cable and found that the fault was still closer to the Bamfield station. We moved over close to the mooring buoy on Seapool Rocks and made another cut into the cable. This time we were able to make contact with Bamfield, but it was getting dark and the sea was rising badly. I can remember distinctly that the cable was lying

on the deck secured by stoppers. All of a sudden there was an extra big sea and the cable ripped out of the stoppers and disappeared overboard. The operation had to be suspended until the next morning when it was finally completed. But if anybody had been in the way of that cable when it whipped out over the bow, that would have been the end of him! Really it was a dangerous operation because the sea had risen quite badly. Making the final splice was very tricky.

When the *Restorer* had finished splicing in a new section of cable after the fault had been cut out, she anchored somewhere near Satellite Pass and sent a boat to report to the station. Berty Baxendale and some of the others from the station came out to see the *Restorer* in a launch that belonged to Richard Clark. It was a fast boat with a secondhand automobile engine. They drew alongside and came aboard. The Captain was so horrified that they had ventured out in such a heavy sea in such an unseaworthy craft that he kept them on board for twelve hours until the sea moderated, then escorted them back into Bamfield.

At that time a new wing was added to the north end of the old Bachelor Quarters. It contained twenty-four single rooms for the additional staff required to man the duplicated system. However, as the number of bachelors decreased over the ensuing years and the number of married men increased, this new wing was later converted into flats for married staff.

CHAPTER
FIFTEEN

# Auckland 1925-1926

FOWLIE WAS TRANSFERRED FROM Bamfield to Auckland in 1925—again on a temporary basis. Describing his experiences there he said:

> I arrived there one evening. One of my daughters had come down with measles on the way down from Canada and we were not allowed to go into any hotel. This was rather a problem. However there was an excellent bond of fellowship between cable people in those days—as there still is. That evening Mr. Blackshaw, the assistant manager of the cable station, came down to the ship to see me and I told him about my problem.
>
> "That's all right," he said. "You and your wife and two girls come along and stay with us while you go house hunting."
>
> Next morning Mrs. Blackshaw arrived with her car and we set out to find a house. We were fortunate. We found an apartment on Edie Avenue, near Remuera Road, which had a lovely view of the sea. The owner had gone to England for six months.
>
> In due course I reported to the office and my training as an engineer commenced. Mr. K. C. Cox was Chief Engineer in charge of the station. He was very nice and tolerant. Fellow trainees, Pace, Peglar and Westwood were very good and helped me with my training.
>
> I enjoyed my time in Auckland. Tam Smales and his wife took us for a picnic in Murawai, a popular resort, and used to invite us to play tennis on their lawn court during weekends.

It was at this time that Fowlie's orbit and mine crossed—with distressing consequences for me.

In 1923 the cableship *Stephan* had laid a cable from Sydney to Southport, Queensland, replacing the old telegraph line. The main trans-Pacific route now became Sydney-Southport-Norfolk Island-

Suva-Fanning Island-Bamfield. All stations between Sydney and Bamfield had been mechanized with automatic relays which superseded the old "gum and punch" method. The Sydney-Auckland cable was now used mainly for intercolonial traffic. It was not mechanized.

The Sydney-Melbourne overland telegraph circuit had been mechanized using automatic transmitters and receivers.

The training school at Southport had been closed and boys were now trained in Sydney.

After being in Sydney for a little over a year, I was transferred to Auckland. I arrived there on my twenty-first birthday.

Arrangements for my lodging had already been made by Bill Stoupe, who had been on Fanning Island with me. I proceeded directly to the place, a private home on Empire Road in Onehunga at the foot of One Tree Hill. My landlord was an old veteran of the Maori Wars in his nineties, who related many tales of the early days in New Zealand. He was married to his second wife, a Maori of middle age, who was an excellent cook. I had the full use of their living room which contained a library of good books and I read avidly.

Much of my spare time was spent roaming the grassy slopes of One Tree Hill, an extinct volcano and once a Maori fortress, with a couple of books under my arm. The summit commanded a panoramic view for miles around, encompassing both shores of the North Island of New Zealand from Manakau Harbour to the Hauraki Gulf.

I joined the Workers' Educational Association and took a weekly course in the basics of architecture which proved to be very interesting. I also attended performances by visiting artists such as Pavlova, the dancer, and Paderewski, the pianist, with the sister of one of my colleagues. Members of the staff also played lawn tennis at the home of one of the supervisors, Tam Smales.

After I had been in Auckland about a year I arranged for my mother to come and stay at a select boarding house nearby for two weeks. She travelled first class on an ocean liner both ways. It was the first and only time she ventured overseas. At that time when ocean liners came into port the passenger list would be published in the newspaper, as news of great interest. Travel was not as common then as it has now become.

Occasionally young bachelors of the cable staff used to get together when off duty and have a good time. This consisted of going from one pub to another in the evening and really enjoying themselves. I dimly remember one occasion when, apparently having exceeded my limits, I passed out. The boys put me in a car and drove me to my lodgings in Empire Road late in the evening. They unlocked the front door with my key, and put me to bed. Strangely, although my eyes were closed, I was fully conscious and knew what was going on, but could not move.

Not long after that I was asked to leave my lodgings. My landlord explained that his wife found it too much work to cope with a lodger.

I found other accommodation near Ellerslie Racecourse with two elderly spinsters who, I think, wanted a lodger for protection. When I first went there they had a round ebony ruler, with a wrist strap attached like a policeman's truncheon, hanging on the hall stand. Shortly after my arrival it disappeared.

Sometimes we were required to work overtime at the office without any advance notice. On one occasion, when on evening duty, I had to work all through the night. When I arrived at my lodgings next morning I explained what had happened, but my explanation was received with disbelieving giggles from the spinsters.

After I had been in Auckland for nearly two years, the main route was interrupted by a break in the cable at Norfolk Island. This meant that all Australian international traffic had to be routed via Auckland on its way to Suva and onwards. As Auckland was not equipped with relays there was only one way to handle this vast amount of traffic—by Kleinschmidt translation. In this system an operator acted as a human relay. The receiving tape passed over a Kleinschmidt keyboard perforator and the operator, reading the signals, translated them onto perforated tape for re-transmission onwards. In this system, errors could be made both in reading the signals and in re-transmitting them. In order to increase the capacity of the cable the speed was raised, increasing the skill and concentration required. Local management decided that Roy Hoskins and myself could handle this situation, so we were each placed on twelve-hour shifts. I was put on the noon to midnight shift, and Hoskins on midnight to noon.

Under the pressure I had to take off my coat, roll up my sleeves, loosen my tie and even take off my wrist watch. For most of the time I was the only one working in the office. The others, with nothing to do, gathered in groups and talked away the time.

The branch engineer, K. C. Cox, used to come into the office in the evening to see how things were going. On one occasion he bent over to look at the signals closely, then looked sideways at me quizzically as if to say, "Are you reading these signals?"

The evening traffic consisted mostly of coded business messages. Errors were traced to the operator who made them. If he made too many he lost his annual increment and seniority. One day as I entered the office, I looked on the error file and saw, to my surprise, that a batch of ten errors had been logged against me. They were all clashes—that is, made by hitting two keys on the keyboard perforator at the same time. I prided myself that I never did this because I could tell when I hit the wrong key. I protested that the errors must be mechanical—that is, made by the perforator—but to no avail. I had no recourse but to accept them and they were logged against me. This so undermined my confidence that my sleep was disturbed by distorted nightmares and I soon became exhausted. The doctor recommended a few days' rest.

Upon returning to the office I thought it better not to take up Kleinschmidt translation again, but the kindly supervisor, Tam Smales, persuaded me that I could do it and that it was the only way to regain my confidence. So, reluctantly, I went back to the job. A few days later I found the same keyboard perforator clashing on the same letters as before—proof that the perforator was responsible for the errors—not me. I had caught it red-handed! The station engineers, K. C. Cox and William Fowlie were called in to witness the incident and agreed that it was the same instrument that had caused the previous errors and accepted them as being due to mechanical failure.

My confidence was restored but it was a long time before the shock and the nightmares faded away. It was not until fifty years later, when we were both retired and Fowlie was my guest at my seaside cottage in Bamfield, that he confessed that he was responsible for my temporary breakdown, having carelessly mis-adjusted the instrument during daily routine maintenance.

Keyboard translation continued for about two more weeks until the cableship had repaired the cable. In that month I received and transmitted over one hundred thousand words without errors. We did not receive any extra allowance for this work—not even an official acknowledgement. It was considered to be all in the line of duty.

CHAPTER
SIXTEEN

# Norfolk Island 1926-1927

SHORTLY AFTER THIS INCIDENT Fowlie was transferred to Norfolk Island in the South Pacific. We did not meet again until fifty years later when, as explained in a later chapter, we met at Annual Veterans' Reunion in Victoria B.C. I invited him to spend a week with me at my summer cottage in Bamfield where I interviewed him about his experiences at various stations in the service. Regarding Norfolk Island, he said:

> I was sent to Norfolk to replace Mr. Lanyon, the station engineer. He was in a hurry to leave, so instead of waiting for the direct boat which was not due to leave for several weeks, I went via Sydney on the ss *Tofua*, leaving my family behind in Auckland. In Sydney I boarded one of Burns Philp's trading schooners which went to the New Hebrides after calling at Norfolk Island. It was a lovely voyage in good tropical weather. En route we called at Lord Howe Island where I went ashore for a few hours. Three days later we arrived at Norfolk Island on a fine moonlight night. We anchored about three quarters of a mile off-shore. I could see date palms and the large stone buildings of the old convict settlement. Next morning boats came out and took us ashore. I was rather surprised that there was no one from the Cable Station to greet me.
>
> Eventually an elderly man came forward and asked me if I was going to the Cable Station. When I told him that I was, he said, "They sent me here to take you out there."

So Fowlie got into the man's sulky. He was so excited that he forgot all about his luggage. He had assumed that it would have been handled by porters just as it was on ocean liners—not realizing that when travelling among the islands, one had to look after his own luggage. The result was that it went on to the New Hebrides and he didn't see it again until the ship returned three weeks later.

In the sulky they jogged along a valley with grass-covered hills on either side. It was a perfect day. Presently they came to a mile-long avenue lined with Norfolk Island pine trees. The road had a soft surface which reminded him of Rotten Row in Hyde Park, London. Then they came to a road lined with lemon trees laden with the yellow fruit. As they jogged along the old man turned to Fowlie and asked, "Are you barren?" Fowlie was nonplussed at first then realized that he wanted to know if he had any children. As the man's daughter was the local school teacher, her salary depended on the number of pupils attending the school. He was pleased to learn that Fowlie's two children would ensure her salary in the future.

They jogged along for another two miles or so and then came to the Cable Station. Fowlie stayed in the Bachelor Quarters while he awaited the arrival of his wife and family who were expected to arrive about three weeks later on a ship coming direct from Auckland.

"Mr. Warner was in charge of the station," he said. Other members of the staff were: Tot Chapman, Elliston, Bill Sanders, Lockley and myself. Sanders, or Sandy as he was called, was very clever. In later years he became divisional Manager in the Persian Gulf for the Company. He had a very distinguished career.

> I was taken over to the house that we were to occupy when my family arrived and we checked the inventory of the furniture which included a set of Mrs. Pott's irons, the kind you heated on the stove. The furniture was old-fashioned and the walls of the various rooms were made of tongue-and-groove wood painted green. The house had a garden surrounded by a hedge. I found out that if I wanted milk, I had to buy a cow. In due course I bought one for five pounds.
>
> The Cable Station was approximately four miles from the place where I had landed, and the inhabitants were scattered over the island. In order to get around, I would have to buy a sulky, horse and harness. Not having been brought up in an agricultural country, I didn't know anything about horses or sulkies or anything of that nature.
>
> It just so happened that around that time the model T Ford had been superseded by the model A in New Zealand and farmers and other people were getting rid of their sulkies, which were a dime a dozen, and buying model A's. So I asked my wife, who was still in Auckland, to buy a sulky and bring it with her to Norfolk Island. She bought a lovely sulky with solid rubber-tired wheels which was the first of its kind on the island and the object of much curiosity.

My wife sailed from Auckland in an awful gale that raged for three days. In due course the ship arrived at Norfolk Island. Now, there are three places where a landing might be possible. The most favoured was at Kings where there was a jetty and boat landing. When I came ashore I just stepped out of a small boat on to a landing, but when she arrived the surf was pounding in there and a landing was impracticable. The ship went around to another place known as Cascade Bay but conditions were equally bad there. There was a third place known as Headstone because two soldiers had been drowned and buried there with two headstones set up as a memorial for them. This place was a last resort.

The Captain had to decide whether to attempt a landing there, or take my wife and two children up to the New Hebrides and land them on the return journey. Well, my wife was a poor sailor and there was no proper accommodation on this small ship for her and her family. One of the officers had given up his cabin for her.

Eventually the Captain decided to make a boat landing at Headstone. It was pouring with rain and there was a big swell coming in. They got into a boat and rode the waves to the landing. Now at Headstone there was a cliff some two hundred feet high. At the bottom of the cliff there was a rocky platform. Ringbolts had been driven into the side of the cliff and a couple of stout ropes were fastened to them. When the lifeboat approached, rising and falling on the swell, two Norfolk Islanders grasped these ropes and leaned far out from the edge of this platform and, as the boat rose on the crest of a wave, the men in the boat held up my wife on the gunwhale of the boat and, as she lifted up her arms, the men ashore grabbed them and pulled her up on to the platform. The eldest child came ashore the same way while the smallest, who was only two-and-a-half years old, was held up by one man and literally thrown into the arms of one of the men hanging onto the ropes. And that's how my family came ashore! It was pouring with rain. We struggled up the steep trail leading to the top of the cliff and I put my wife and children into a borrowed sulky. Dripping wet, my wife asked, "What on earth will the people on the station think of me in this outfit after three days of seasickness and wet through in the pouring rain?"

I didn't know anything about a horse but managed to drive the two miles to the station without any trouble. I suppose the other wives on the station were watching our arrival from their windows, but by the time we got there we were past caring. We entered our house and I got a fire going and we thawed out.

The handyman on the station came every morning to milk the cow for us. I don't know if this was the usual method of milking on most farms, but in those days in England they just sat on a three-legged stool and did things to the cow and the milk came out. On Norfolk, before you could do that you tied up the cow, then got one of its hind legs and tied that up too—so that it couldn't kick you.

Groceries came out once a week from a store in the middle of the island. There was only one motor-car on the island in my time and the owner had to give twenty-four hours notice before he could drive it on the roads. He had to do this because the horses on the island had never seen a car before and took fright when they saw this one coming.

There was a tree at the crossroads in the middle of the island that served as a notice board, and the owner had to post a notice stating that on such and such a day he would be using his car, and you were advised to use due caution if you heard the car coming in the distance; you dismounted and stood by your horse's head until the car passed.

When I got the sulky, I found that I needed a harness. Well, it took about a month to acquire one. Then I had a disagreement over the price for the horse. I had intended to purchase one from my predecessor but someone told me that it was not much good—that it was broken-winded. Not knowing what that meant I asked the previous owner what was the pedigree of the horse. I didn't know that a pedigree was something that only thoroughbreds had. Well—it became a great joke on the island about me asking for the horse's pedigree! In the meantime I managed to hire a horse when we wanted to go downtown, which was about four miles away. It really wasn't a town. There were just a few small farms.

Down near the main landing place there were the remains of the old convict settlement. The convicts were there from 1790 to 1850, when the authorities decided to stop sending them there and they were taken away from Norfolk Island. Around this time, Queen Victoria made a gift of the island to the descendants of the mutineers of the *Bounty*, who were having difficulties with over-population on Pitcairn Island.

I liked the Norfolk Islanders very much. They were very good to us. Among themselves they spoke a language that was unintelligible to me. It was composed of a mixture of an ancient west of England dialect and words from the Tahitian language. When two islanders got together they spoke their own language, but when they came to me, they spoke English.

We had a woman who came out on a pony to do our washing. Her name was Quintal—the name of one of the original mutineers of the *Bounty*. Just along the road was another descendant who had a banana plantation. Getting firewood was a problem. There wasn't much available and we did not have an oil stove. The wood was so hard that it was difficult to split. My wife complained because we couldn't get enough to get a decent fire going in the kitchen stove. Then there were those Mrs. Pott's irons she had to use for ironing the clothes. Although she had never baked bread she had to make her own. A neighbour came over with a bottle containing a preparation for bread-making. The cork had to be kept tied tightly after each use. One night she forgot to tie it on and, in the middle of the night, the cork blew out and splattered the contents on the ceiling, staining it. She was very distressed. Eventually she learned to make her own bread.

But the cow and I never became friends. She used to chase me when she got annoyed!

On the whole we got along very well on the station. I did not see eye-to-eye with the manager, Mr. Warner, but I don't blame him. I think I must have been a bit of a pain in the neck to him. The station was situated on the edge of a cliff. The cliff dropped off almost perpendicularly to a patch of sand in Anson Bay where the cables landed. There was a steep path leading up this precipitous cliff. Once a month we had to go down to the cable hut and inspect the cable terminals. To return to the top of the cliff up the zigzag path took about twenty minutes. The landline connecting the cable terminals to the office was buried in a trench all the way up the cliff into the office. All three cables came up that way.

The office was a wooden building with the manager's house next door. Farther down was the old quarters building which used to accommodate the bachelors on the staff. At one time, during the "gum and punch" era, there used to be a fairly large staff. Messages were received from Suva and relayed, by human relay, on to Southport. Also there was another cable from Norfolk Island to Auckland.

By the time I had been transferred to Norfolk Island (1926) the cable had been duplicated between Suva and Bamfield and the island had been converted into an automatic relay station. All the apparatus there had been invented, patented and manufactured in England for K. C. Cox, a very clever P.C.B. engineer. These instruments received the signals from Suva, or from Southport, depending on which direction they were going, clearing up all their imperfections due to the characteristics of the cable and re-transmitted a near-perfect signal to the next station on the line. One of the chief features was the Cox magnifier which took the incoming signals and increased their amplitude to such an extent that they could operate the relay. Before this, the amplitude of the signals was too small to do anything of this nature.

Cox also invented an interpolator which regenerated a near perfect signal for onward transmission. Another invention of his bypassed all but the most severe outbursts of electrical disturbances due to the aurora. Sometimes these magnetic storms, or earth currents, were so severe on the old Bamfield-Fanning cable that it would be unworkable for twenty-four hours. He also invented a reperforator which not only received the signals, but also transformed them into a perforated tape which could be stored for onward transmission by an automatic transmitter.

It was a wonderful system. I thought I would be in charge of electrical matters on Norfolk, but really it was Sanders who did all the work and showed me what to do.

There was only one person on duty at a time in the operating room, and there were three shifts of eight hours each. The assistant engineers covered the day and evening shifts. We didn't have to do the night shift, but were subject to call if anything went wrong. The manager came on at ten o'clock in the morning and again in the afternoon. He was very efficient and kept a pretty close watch on us.

We had a power plant that generated direct current for lights on the station. The engine house, which was about a hundred yards from the main station, contained a gasoline driven dynamo with an elaborate switchboard. There were two large hundred-volt batteries with sixty cells each in this engine house. Just at this time the two batteries were being replaced. The new batteries had been landed and placed in position. After each battery was filled with sulphuric acid, it had to be given an initial charge which continued for twenty-four hours without interruption. After that the cells were given a rest and then charging could continue until the cells had received their full charge.

I was placed in charge of the gasoline engine and charging apparatus from eight o'clock in the evening until eight the next morning. It was a hot, wet, very humid Sunday night. I went on duty and relieved the branch engineer, Mr. Bailey, who rode off on his horse to his home at the other end of the island, leaving me alone on the station. I had to take the specific gravity of each cell and note its temperature every hour. The battery room had no ventilation and soon the acid content of the air was very high.

Next morning, while Fowlie was making final checks before being relieved by Sanders, there was a violent explosion in the battery room due to a leakage of gasoline ignited by contact with the hot exhaust pipe from the engine. Fowlie escaped unharmed but the battery room and its contents were completely destroyed. Although physically unharmed, Fowlie suffered from shock which lasted for several days. In his own words he said:

I went through hell. Bill Sanders, who was my junior, took charge and helped me. He did wonderful work. I was pretty well shell-shocked. Our power was gone and we had no electric lights, but somehow Sandy did wonders. Mr. Cox's magnifier required an intense electric beam, so Sandy rigged up an apparatus so that it would work with an oil lamp. We kept those cables working, perhaps not at their full potential, but working remarkably well throughout that difficult period.

After a while I began to come out of my shell-shock. I enjoyed the last part of my time there. The climate was really lovely. It was a bit chilly in the middle of winter but it wasn't really cold. The main trouble was the high humidity. The cliff near the office was three hundred and twenty-five feet high from the sandy beach below and the spindrift from the waves breaking on the rocks below came up over the top of the cliff, saturating the air with salty spray. Working in the office was very difficult because the moisture would cover all the instruments. Before you could do a routine test, you had to shut the door to the test room, light oil lamps and keep them lit for twelve hours until the place became reasonably dry before you could conduct tests.

The houses were affected the same way. If you leaned against a wall of your house, which was of clapboard construction, your shirt would get wet through.

After his term on Norfolk Island, Fowlie was transferred back to Auckland in 1927 for six months or so and then on to Fanning Island.

CHAPTER SEVENTEEN

# Fanning Island 1928

En route to Fanning Island Fowlie spent two weeks in Suva, Fiji, awaiting the cargo ship *Wairuna*, bound for Fanning.

We spent a happy time in Suva until my youngest daughter came down with dengue fever. Now there was trouble! A crisis was expected and the doctor watched her closely. When our ship came into port he said he did not know if she would be well enough to leave.

"If your daughter's crisis comes before midnight, and she gets over it," he said, "I will let you go."

We waited anxiously all that evening. Eventually the doctor decided that she had weathered the crisis. Next morning we were taken aboard, together with a dozen chickens we were taking with us. The *Wairuna* was formerly a German ship that was given to the Allies as part of the compensation for the 1914-1918 world war. She was now on the Sydney-Vancouver run calling at Fanning with provisions, mail and passengers every two or three months.

It was a delightful voyage with good weather all the way. The Captain took a fancy to my little four-year-old daughter, and took her for walks around the deck. Now, when we were in Fiji, we had gone fishing several times and trolled a line over the stern of the boat. So, one morning, while walking with the Captain, they came to the stern of the ship and saw the log line trailing behind to show how fast the ship was travelling. The Captain tried to explain this to my daughter. She doubted his explanation and, unable to control herself said, "Don't be silly! That's for fishing!" Well, her remarks were overheard by a sailor who happened to be in a small housing containing the auxiliary steering wheel. The joke was told and retold around the ship for days. "Don't be silly!" they said.

Eventually we arrived at Fanning and were landed on the beach by a surf boat and taken to our house. Every morning a native shinnied up a coconut palm and filled our refrigerator with nice fresh coconuts full of coconut milk which made a delicious drink—especially with the addition of a little gin. The white "meat" on the inside of the nut was more of a jelly when fresh. The natives used to feed it to their babies. When dried it is called copra, the main source of revenue for the local copra plantation.

The cable station had a strip of land from sea to lagoon about half a mile wide. The rest of the island belonged to Fanning Island Plantations Ltd., owners of the copra plantation.

The Bamfield-Fanning-Suva section of the cable system had been duplicated two years previously, providing nearly ten times the capacity of the old, original cable. Relays had been installed on both cables, replacing operators. These instruments received the signal, re-shaped them into near perfect signals, and re-transmitted them on to the next station. They were fully automatic. All that was required of the attendant was to keep a watch on electrical adjustments and see that everything was operating satisfactorily.

There were three operators and three assistant engineers, of which I was one. In the office, lizards about four inches long climbed everywhere. Sometimes one would climb on the large, white, translucent shades over the light bulbs and cast a long shadow on the ceiling that looked like a crocodile.

In my spare time I ran the post office. It was opened only for two or three nights before a ship was expected. I was very busy for those few days because everyone had written reams of letters and there were lots of money-orders to be issued to people who had ordered things to come by the next boat. Also I made up private orders to come by the trading schooner, *Doris Crane*, from San Francisco. These consisted of staff orders for anything from clothing to cameras, typewriters, etc. Fanning Island Ltd.'s office in San Francisco would hand these orders over to Wightman and Crane Ltd., to make up and ship by the schooner.

Besides the post office, I also ran the local bank which, like the post office, came under the auspices of the New Zealand government. The banking system was unique. The manager paid the staff in New Zealand currency. Then they came to me and deposited most of the money in a post office savings account. I placed these amounts to each man's credit, then took all the money back to the manager. He sent a cablegram to the post office in Auckland, where each man's account was credited with this amount. So the manager got back all the currency he had distributed, and the same currency was recirculated next pay day. Before my time the staff had been paid in gold sovereigns. Long after gold had ceased to be the currency anywhere else, it was still being used on Fanning as the method of payment.

There were large numbers of crabs on the island. Crabs of all shapes and sizes. The most prevalent were the land crabs, which were about eight inches across with two claws, one larger than the other. They scampered all over the station compound and pathways and burrowed in the earth like rabbits. They were not afraid of people. Usually they would scamper out of the way but in the dark you couldn't avoid treading on them. Their eyes were elevated on little stalks half an inch high sticking up out of their heads. They watched you and would wave their claws as if they were going to attack you, but they didn't—they kept out of your way. My youngest daughter was terrified of them at first so I gave her a stick and told her to chase them out of her way. From then on she had no fear of them.

While I was there we fitted out the tennis court with lights for night tennis. Playing tennis under the lights when the temperature was in the mid-seventies was very pleasant. The temperature didn't vary much on Fanning. Being almost on the equator, nights and days were almost equal, varying only about twenty minutes between summer and winter. The sun set about six o'clock every day and the evenings were always nice and warm.

We had a doctor on the station who performed surgery when necessary. He was a dear old man who had been pensioned off. I had to visit him when I had a severe toothache. Joe Nevitt, one of our staff, came over with me. He, and somebody else, held my arms and pinioned them behind my back while the doctor got busy with a pair of forceps. He yanked and pulled until something gave way. But all he had done was to pull out a large filling in the tooth and from then on it ached more than ever. By this time I had had enough, so he gave me a stiff dose of overproof whiskey which, he said, should give me relief for a couple of days. The toothache died down after three or four days. It would be another year before I could leave Fanning and have the tooth extracted.

I left Fanning in 1929 on the *Wairuna* again. I travelled across Canada making stopoffs at various places, then sailed on the *Tuscania* across the Atlantic and landed in London on five months' leave.

CHAPTER
EIGHTEEN

# Duplication of Pacific Cable

As early as 1912 the Pacific Cable Board considered duplicating the Suva-Fanning Island-Bamfield section of the Pacific cable in response to increasing competition from the recently invented system of wireless telegraphy. Their concern was evidenced by the following paragraph in the *Port Alberni News* on 21 September, 1912.

> The P.C.B., instead of duplicating the existing cable to Australia, is considering the installation of wireless telegraphy using high power. The opinions of experts have been sought on the matter.

On 20 June, 1914, the newspaper reported:

> The P.C.B. has decided to lay a second cable from Bamfield to Port Alberni, a trenched cable from Port Alberni to Parksville, and a submarine cable from Parksville to Nanaimo, the latter to make connection with the cables from Nanaimo to Vancouver. This will give the P.C.B. an all-cable route direct from Australia and New Zealand to Vancouver. The cable ship [*Iris*] left Auckland on June 16 for Bamfield. She will probably land the new cable at Port Alberni on July 29.

However the cable between Port Alberni and Parksville was not entrenched until about fifty years later.

In 1923 the P.C.B. again considered duplication of the cable between Suva and Bamfield. Before proceeding, experiments with long-distance radio were conducted between Bamfield and Sydney. A team of radio technicians arrived at Pachena Point lighthouse, on the west coast near Bamfield, in September of 1923 and set up equipment for high-speed transmission and reception of radio telegraph signals. Contacts with Honolulu and Sydney were established but proved to be unreliable, mainly because at that time, the system of beaming signals by means of special antennae had not been invented—or at least not used by those who experimented.

After the failure of these experiments the duplication of the cable was decided upon. On October 28th, 1926, the 9,000 ton cableship *Dominia* which, like her predecessor the *Colonia*, had been specially constructed for the Pacific cable project, sailed from Bamfield and began paying out cable at the rate of ten knots. The cable, a new type described as "loaded" had a thin wire permalloy wound spirally round the copper core, which increased the inductance of the cable in order to neutralize the effect of the capacitance. This permitted a remarkable increase in speed, six times that of the original cable, but only under simplex conditions: one direction at a time.

After travelling an average distance of two hundred miles each day, the *Dominia* arrived at Fanning Island on the 14 November and buoyed the end of the long cable. Another cableship, *Faraday*, was duplicating the section between Suva and Fanning Island.

When speed trials were conducted on the new cable, they proved that the expectations of the design engineers had been correct: a simplex speed of 150 words (750 letters) per minute was attained, making it not only the longest cable in the world but also the fastest in the Commonwealth.

While the cable was being duplicated, a new concrete three-storied building was erected in Bamfield in front of the Bachelors' Quarters. The austere edifice, perched on the edge of a cliff overlooking Barkley Sound, resembled a Tibetan lamasery and was referred to by unhappy bachelors on the staff as "Alcatraz."

Although the new cable had only one copper core, three channels were superimposed on it by means of synchronization—each worked at a speed of 250 letters per minute. One went to Sydney, another to Auckland and the third to Suva. The signals on this cable, despite its long length, were easy to read, every dot and dash being clearly defined.

With the speedy service that resulted from the duplication, the Pacific Cable Board captured an increased amount of traffic from their competitors, Amalgamated Wireless of Australia. However, shortly afterwards, the invention and introduction of the beam radio antenna in 1927 so improved transmission and reception of long-distance radio signals, that the beam radio—which offered a cheaper rate—took the business back again. Beam antennae concentrated the signals like a searchlight instead of broadcasting them, thus resulting in a much stronger signal.

CHAPTER NINETEEN

# Crisis!

WITHIN SIX MONTHS OF ITS introduction in 1927 beam radio captured a large percentage of Eastern Extension's overseas traffic and more than half of the Pacific Cable Board's. Cablegrams from England to Australia cost two shillings (twenty-four pence) a word for full rate messages versus the Beam's four pence. The competition was more than the cable companies could stand. They appealed to the Conservative government of the United Kingdom, which owned and operated the beam services, to save them from total collapse. The situation was desperate and demanded immediate attention.

The reason for the disparity in rates was obvious. Cable companies had to staff and maintain numerous relay stations between the terminals of London and Sydney, as well as the intervening cables and a fleet of cableships, whereas the beam radio stations communicated directly between terminals.

In order to avoid a rate war, the government invited competing companies to meet and discuss the various problems facing them. Following the meeting the government called an Imperial Wireless and Cable conference to discuss the problems and to make recommendations. It recommended the formation of a Merger Company and a Communications Company. The latter would purchase radio and cable stations, cables, cableships, etc. It would buy the Pacific Cable Board and manage the entire Imperial cable and wireless service. Its name, Imperial and International Communications Ltd., was later changed to Cable and Wireless Ltd.

It was stipulated that before I.I.C. could take over, it was obliged to continue to employ all existing staffs under terms no less favourable than those they already enjoyed.

The reasons for the government's acquiescence to the merger were twofold. First to avoid financial losses by all concerned; second to maintain a high level of security. Cables were a reliable secret means of communications—especially in wartime—and must be maintained at all costs. Radio, on the other hand, could be intercepted by the enemy and at that time was subject to lengthy periods of disruption by fading and blackouts caused by magnetic storms. When this happened, all beam traffic had to be diverted to the cables for transmission. On the odd occasion when the cable service was interrupted by a fault or breakage, the cables diverted to the beam.

On August 2, 1928, the government introduced an Imperial Telegraphs Bill to authorize the sale of the Pacific Cable Board for £517,000. The P.C.B. had an outstanding debt of £1,233,000, the remainder of the original loan of £2,000,000.

The sale was opposed by Ramsay MacDonald, leader of the Labour Party opposition, on the grounds that it sacrificed public utility to private gain by disposing of valuable State undertakings to private interests. "This is not a little pettifogging, superficial thing," he said. "It is one of the biggest pieces of national policy, of national business that has ever been brought before this House, at any rate in my time." One member thought it was as absurd to hand the Government cable system over to private enterprise as it would be to hand over the Navy.

The Bill received royal assent on 5 February, 1929.

This was the time of the Great Depression. In a cost-saving measure, all Cable and Wireless staff were required to take a ten percent reduction in salary. Cable and Wireless also introduced a voluntary retirement scheme which enabled many to accept early retirement.

CHAPTER
TWENTY

# Transfer to Bamfield 1930

AFTER BEING STATIONED IN Sydney office for two years, I volunteered for transfer to Bamfield. I sailed on the old *Niagara* in early March, 1930. Travelling first class on an ocean-going liner was my idea of the good life, especially since my passage, including wine and tipping allowances, was paid by the Company. First class passengers dressed for dinner and were fed unusual delicacies such as "milk-fed suckling piglets" and many other tasty morsels. I had my first taste of the famous American dish of strawberry shortcake and cream.

In those days the Captain of the ship inspected each cabin weekly. It was not very thorough—he merely looked inside the cabin perfunctorily. It was an opportunity for anyone who had a complaint to make it directly to the Captain. On board with me were Jack Turnbull, another cable operator bound for Bamfield, and his wife, Moragh. She boarded in Auckland where she had been visiting friends. We all sat at the same dining room table, together with a young girl bound for Honolulu. In those days ships were not equipped with air conditioners or stabilizers, and dancing on deck in the evenings was a real hazard. One moment you were dancing uphill as the ship's prow rose, and the next one you were sliding downhill as it plunged into a trough. When it listed to port or starboard, it was as much as you could do to avoid being dumped into the scuppers.

It was early spring weather when I arrived in Victoria, en route to Bamfield. The weather was clear, cold and sparkling. It was an entirely new experience for me, accustomed as I was to warmer climates. It was exhilarating. I disembarked from the *Niagara* and stayed overnight in the Empress Hotel awaiting the arrival of the small coastal steamer

*Princess Maquinna* to take me to Bamfield. I was invited to dinner by relatives of my fellow passengers, the Turnbulls, and made lasting friendships with them and other members of the Jameson clan.

The following evening I boarded the *Princess Maquinna* and made the overnight journey to Bamfield. I was greeted by members of the staff whom I had known in Sydney, and taken to my room in the Bachelor Quarters. There were about twenty bachelors, one of whom was elected Messman to make up the menus and order foodstuffs from the wholesalers. They paid only for their food, accommodation being free.

For outdoor recreation there were various sports—tennis on concrete tennis courts, hiking on forest trails, boating, swimming on ocean beaches, salmon fishing, hunting both game and duck, and photography.

For indoor recreation there was a dance hall equipped with a stage for amateur theatricals and concerts and a cinema booth for movies. Life on the station was like living at a country club.

Within an hour of my arrival I had changed clothes and was hiking through the forest on the West Coast Lifesaving Trail which skirted the coast from Bamfield to Port Renfrew, forty miles to the south. The primeval forest was impressive. The area, at that time, being one of the most thickly wooded in the world, so dense that only the smallest birds could penetrate it. The silence was sepulchral, so different to the lively Australian bushland, always filled with the sound of birds. The sweet multi-scented air of the forest mingled with the salty tang of the ocean, producing an exhilarating atmosphere.

As yet no system of automatic working had been devised to relay signals from the world's longest cable into one of the world's longest landlines across Canada. Both were over four thousand miles long. The "gum and punch" system was still in use with operators acting as human relays.

There was only a "physical" telegraph circuit across Canada to Halifax, 4,000 miles away, maintained by the Canadian Pacific Telegraphs. It is a wonder that it worked as well as it did. Often we found that we had been sending into "thin air" having lost contact with Halifax.

During periods of magnetic storms the C.P.T. was unable to keep fuses in the circuit—the electrical impulses generated by the storm blew the fuses as fast as they were replaced. Likewise the original slow cable would be unworkable due to continuous alternating current of reversals generated in the cable by any severe magnetic storm. The new "loaded" cable was not affected.

The traffic pattern was curiously tidal, with the outbound messages from Europe and America flowing through Bamfield from 9 a.m. to 9 p.m.; then after a slack period, it would reverse and flow in the opposite direction from 9 p.m. to 9 a.m. The new simplex cable was switched, or changed over, for use in whichever direction the traffic was flowing, the slow cable being used in the other direction since there would be little or no traffic.

Shortly before my arrival the Pacific Cable Board had been taken over by Cable & Wireless Ltd. Australians and New Zealanders stationed in Canada at that time were given the option of accepting Canadian domicile or "their employment would not be guaranteed." I accepted but some did not. They requested and received repatriation back to their domicile.

This ended the Pacific Cable Board era.

At this time the Australian Overseas Telecommunications Commission assumed control of the Pacific island stations of Suva, Fiji and Fanning Island, providing staff and maintenance.

## Social Activities

Apart from the cable station, Bamfield was just a small fishing village with a population of about two hundred people. Being isolated from the outside world by dense forest and ocean waters, it had never been "developed." Although it had numerous means of outdoor recreation, it had none for indoor entertainment—except a community hall. The cable station had a dance hall with music supplied by a five-piece orchestra composed of members of the staff. On Saturday nights, the staff also provided movie shows which were open to the public.

Highlights of the year were the annual tennis dance at the conclusion of summer activities, and the New Year's dance, to which people were invited from all around Barkley Sound and even as far away as Port Alberni. They travelled for hours by boat, danced all night, and travelled back again next day. The weather at this time of the year was usually wet and stormy, so that ladies from across the Inlet had to protect their evening gowns by wearing slickers, sou'-easters and gum boots. When, after being ferried across the stormy waters of Bamfield Inlet, they arrived at the dance hall, or a friend's house on the station, each emerged from her rain clothes like a butterfly from a chrysalis, resplendent in her finery.

During the Christmas season the bachelors entertained the married staff and their wives and friends to a sumptuous turkey dinner with all the trimmings, followed by a dance in the library and all-night revelry.

On Christmas night the bachelors were entertained by the married staff, each married man inviting one or two bachelors to a home-cooked dinner. After dinner, the custom was to go from house to house singing Christmas carols. After being liberally entertained at one house, the whole party would wend its way to the next, singing and frolicking. Even after daylight, men in tuxedos could still be seen wandering about the station compound.

During the summer, Canadian Pacific coastal steamers, *Princess Maquinna* and *Princess Norah* called at Bamfield every ten days on the up-Island trip to Port Alice, and again on the return journey four days later. The trip was very popular and the ships were always filled with tourists. When the ship called in the morning on her way up-Island, bachelors met the boat and eyed the female passengers with a view to making dates on the return journey. It was simple. All they had to do was to offer to show them through the cable station—which everyone wanted to do—and the groundwork was laid for the return visit.

When the ship returned, she laid in port overnight and the passengers were entertained by the cable staff at a dance in their dance hall until the ship sailed in the early hours of the morning.

This was the way I met my future wife.

## Snowstorms in the Rocky Mountains

In the winter of 1934-35 there was an unusually heavy snowfall in the Canadian Rocky Mountains. It was so heavy that railway snow-ploughs were unable to cope with it. Trains were stalled on both sides of the mountains, about five miles apart. Since there was no microwave system across Canada in those days, all communications went by landline and were disrupted by the snowfall. The Canadian Pacific Railway Co., in Vancouver, wishing to send a message to trains on the other side of the Rockies, asked our Vancouver cable office if they could send a message around the world to the other side of the Rockies. This was done, a flash message being relayed across the Pacific Ocean to Australia, across Australia, the Indian Ocean, Africa, the Mediterranean and on to London—thence across the Atlantic Ocean to Halifax and across Canada to the Rockies—all in record time.

On a similar occasion, Cable and Wireless Ltd., wishing to promote the introduction of a new cheap rate service, arranged for the Prince of Wales to come to their London office and personally send a message around the world. In order to facilitate the transmission and reduce the time taken to go around the world, all cables in the world-wide system were physically linked together to avoid time lost in handling. There was only one place where this could not be done: Bamfield. Owing to the long landline across Canada (4,000 miles) and the long cable, also over 4,000 miles, to Fanning Island, the state of the art at that time did not permit the use of relays, nor could the two be linked together. There was only one solution to the problem—a human relay using Kleinschmidt keyboard translation.

Being on duty at the time, I was assigned to the operation. I felt a bit nervous at first—as did the other men—but someone cracked a joke which relieved the tension. Once the message started I rose to the challenge and had no trouble re-transmitting the thirty word message which went around the world in a matter of minutes.

Sometimes during this period the old, slow Bamfield-Fanning Island cable was interrupted. Engineers made tests to ascertain what kind of fault it was and where it was located. A number of tests were made but each one gave a different reading of capacitance and resistance. A single reading should have indicated how far away the fault lay, and what kind of fault it was—partial or complete break.

Nonplussed Bamfield engineers reported the information to head office in London who then instructed the cable engineer on the cable-ship *Restorer*, based in Victoria, B.C., to go to Bamfield and test the cable. His readings varied as much as those of the Bamfield engineers, so he recommended to London that an average of the readings be taken and a new piece of cable be spliced in there. There was silence from London for several days. Then Bamfield was instructed to "cut back the cable-head six feet and report." The cable head was where the cable entered the office building and made a wide semi-circle bend before being connected to the cable terminals. These instructions were followed and it was found that the guttapercha insulation around the old cable had perished, allowing temperature variations to affect the cable. The fault was in the office building!

At times the weather in Bamfield was wet, stormy and humid. Storm clouds which had absorbed heat and moisture from the Hawaiian Islands were blown across the North Pacific to the west coast of Vancouver Island where, upon encountering the coastal mountains, they were forced to dump their contents in tropical downpours sometimes exceeding five inches in a day. The humidity was so great that beads of moisture formed on all the instruments rendering them useless for several hours. Hair-driers, borrowed from some of the ladies, had to be used to dry them out.

We earned a month's leave every year. This could be saved until we had accumulated five months, then we were entitled to long leave, or furlough. Our first class passage back to our domicile would then be paid by the Company, and direct travelling time would not be included in our leave. We could travel any route we wished, the cost of the direct passage being handed to us in advance. I chose to travel the long way around the world via Europe on the Classical Tour, spending two months travelling and the remainder in Australia. If we did this, we also received a free train passage across Canada from the Canadian Pacific Railway Company, with whom we were associated. So, in 1934 I made the classical tour of Europe, returning on the P. and O. liner *Strathnaver*, through the Suez Canal and the Indian Ocean en route to Sydney, calling at Adelaide on the way.

Upon arrival in Adelaide I was surprised to receive a visit from a cousin, Lal, whom I had never heard of before. An employee of the branch office of the P & O Steamship Company, in Adelaide, he had seen my name on the passenger list. As my first names were the same as my father's, he had assumed that I must be his son. I was delighted to meet him. He took me to see the old family mansion "Hawthorne Bank" on the corner of Kermode Street and Avenue Road. It was now St. Peter's School for Girls, run by the Sisters of the Church in Adelaide.

He introduced me to the Sister who answered the door as his cousin from Canada. The Sister, who had known my father when he was organist at the cathedral, said she saw the resemblance and offered to show us around what used to be our family's home. She was fully acquainted with its history and knew who had occupied each bedroom saying this was uncle so-and-so's room and that was aunt so-and-so's and this was the chapel. Our grandfather, William David Scott, who had held the administrative position of Master of the Supreme Court, had had a large family of six boys and four girls.

According to the "History of St. Peter's Girls' School" by Mary M. Frost:

> Mrs. Scott's house "Hawthorn Bank" came to the notice of the Sisters, who thought it would be more suitable for their permanent home than the very small Keble Cottage (they presently occupied). At the beginning of December it was bought to serve as the Community House in Adelaide. It was not yet possible for the Sisters to move into it, but after they did so, it remained for more than half a century the home of the Sisters in this State and the heart of St. Peter's Collegiate Girls' School.

Lal then took me to see the old home of our Great-Grandfather, Captain William Scott, who had sailed out to South Australia from England in his own sailing ship, *Canton*, in 1839. Active in community affairs, he became a member of the State Legislature and was given a grant of eighty acres in 1841. He dismantled his ship and used the timbers to construct his home, which he called "Brookside." The ship's masts were cut to size and used as verandah posts. The large drawing room, with its imported marble mantlepieces, was the scene of many social occasions when Captain Scott entertained Legislative Council members and their wives, dancing the quadrille and the lancers in an atmosphere of candlelight and wine.

One of the original grape growers in the State, Captain Scott had his own vineyard and winery, as well as 2,000 sheep, 16 cattle, six acres of wheat and three of barley. The Scotts had four sons and three daughters.

Lal told me that Scott's wife, Mary, was very pretty but quick tempered. She retained her temper into old age and his father and uncles were always scared of her. She was also courageous—sometimes being left alone with her children on the isolated station for days on end with no possibility of help in case of emergency.

At the time of my visit, the owner, Joseph Guidi, was operating the property as a stud farm for breeding trotting horses. I knocked on the door, and when he answered, explained that I was from Canada and was a descendant of the original owner, hoping that he would invite me in to see the interior of the house. Disappointed that he did not do so, I exchanged pleasantries, and departed.

When I revisited Adelaide many years later, I could not remember the address of "Hawthorne Bank." A nephew drove me around the city looking for the old home, but we could not find it. It had been demolished to make way for a hospital.

"Brookside" was still standing, having been declared a National Heritage site.

CHAPTER TWENTY-ONE

# Sydney and Suva 1934-1935

UPON ARRIVING IN SYDNEY, I STILL HAD three months' leave remaining. After my furlough expired, I reported for duty. The two systems, Via Eastern and Pacific Cable Board had been combined in one building in Spring Street and were now known as Cable and Wireless Ltd. Gradually the P.C.B. men lost their identity (of which they were so proud) and were absorbed by their competitors.

I applied for transfer back to Bamfield, and after only two weeks in the Sydney office, my application was granted—provided that I agreed to serve temporarily in Suva for six months.

In the early days Suva was the station to which most probationary operators were sent on their first assignment overseas. Bachelors were accommodated in the Bachelors' Quarters situated on a hill overlooking the harbour, and were just across the street from Government House. Being a Crown Colony, Fiji was governed by a governor sent out from Britain as the King's (or Queen's) representative.

Since the advent of the new cable across the Pacific, Suva had become an automatic relay station with a minimum of staff, so the Quarters, being no longer required, had been sold to private interests. It was being run—of all things—as a temperance hotel, having been anything but temperate when bachelors lived there. This was where I stayed while in Suva.

The Grand Pacific Hotel, that grand old dame of the South Seas, was the centre of social activities. Located on the waterfront with a beautiful view of the harbour, it was surrounded by palm trees—so much so that it was impossible to get an unobstructed view of the hotel. Its outdoor swimming pool of skyblue water was ringed by palm trees in a romantic setting.

Dances at the G.P.H. were very formal affairs, tuxedos with starched shirt and collar with a black bow tie being obligatory. These were the days before air-conditioning. It was so hot dancing in that humid atmosphere that the starched collar soon became limp and sodden. One had to carry spares.

It was the custom to go to the G.P.H. before dinner for a social drink. One or two would soon be joined by others until, before long, there would be quite a large group. Woe betide any young operator who got caught in such a situation, for the cost of a round of drinks would surely break him! One discreetly left before that happened.

Being a relay station there were no operators, only relay attendants who responded when lights flashed and bells rang indicating that some piece of apparatus was malfunctioning.

One evening while I was on duty, the relay system broke down. Frank McCay, the station engineer, said it would probably take a couple of hours before repairs could be made. This meant that all international traffic would have to be diverted to our ex-competitors, Via Eastern, for transmission and delivery—unless a temporary circuit could be rigged up using Kleinschmidt translation. Asked by McCay if I thought I could handle it, I replied that, having done the same thing in Auckland a few years ago, I did not see why not. I was nervous at first but soon settled down and even enjoyed the challenge.

Two months before I was due to leave, I contracted dysentery. The Company's doctor told me to go to bed for a week and gave me a bottle of patent medicine to take. It did no good. When I reported for duty the manager could see that I was not fit and recommended that I be transferred to Bamfield on the next steamer.

Fred Studman, of Auckland, had happier memories of Suva. Describing his recollections he said:

> I was transferred to Suva in 1924. As the Bachelor Quarters was fully occupied, I was accommodated in the "Waimanu" on Waimanu Road, managed by Mr. and Mrs. Newton. There were approximately twenty guests, mainly school teachers, bank clerks and others on a two year assignment with various businesses around town.
>
> Suva was a pleasant place to spend one's first overseas transfer. Working hours were short, six a day, and there was plenty of sport. We fielded a top grade rugby team composed mainly of staff and were members of the local tennis club.

Mr. Plunkett, a well-respected and popular English gentleman, was superintendent of the station. Practically all of the staff were from Auckland and Sydney on their first overseas assignment.

Quite often, after coming off night duty, the staff would collect a couple of loaves of bread, some butter and tins of salmon, and spend the day picnicing up one of the rivers around the harbour. They would return later in the day to catch some sleep before going on night duty again.

It was the usual practice for staff going on duty from the Bachelor Quarters, to hire a taxi at sixpence a head. Then the staff being relieved would take the same taxi back to the Quarters. The regular taxi driver was an East Indian named Rostam Singh, familiarly known as "Rooster." On a wet day he would take as many as he could squeeze into his cab. On a hot, sticky evening it was not uncommon for members of the staff to climb over the wall of the old Suva Baths nearby for a free cool off before going home.

I was a member of a six-piece dance band as pianist. As we were the only musical combination at that time, we played for all the dances held at either the Grand Pacific Hotel, or the Town Hall. Most of the sports clubs held an annual ball, the most popular of which was the Bowlers' Ball at the G.P.H. The social highlight of the year was the Government House Ball, so everyone made sure that their name was on the visitor's register at Government House. Sir Eyre Huston, the Governor, made a point of personally meeting and thanking members of the band. Sometimes I would have to pay another member of the staff to work for me if a change of duties could not be arranged.

In 1926 the Auckland University Football team visited Suva. The first test match resulted in an 8 all draw. In the second test the visitors won 10 to 3. The final test was won by Fiji 19 to 18, by which time the visitors were probably overwhelmed by hospitality. Our band played for a ball given in the Cable Bachelor Quarters in honour of the visiting team. It was reported by the press that "All of Suva's younger set were gathered there." It was also reported that "The Suvaneer band supplied excellent music and was enthusiastically recalled time and time again." A good time was had by all.

Once I attended a horse race meeting on property out near Suva Point owned by then Mrs. Alport Barker. The so-called jockeys were mainly Fijian and East Indian riders and what they lacked in horsemanship was more than made up by enthusiasm. The local newspaper reported that "In one race recently five horses started. One ran into the bush, two threw their riders and the most scrubby horse of the lot won by several lengths amidst dismayed silence. No one had backed it."

Another facet of life in Suva was described by John Lee, also of Auckland:

The Bachelor Quarters of the Pacific Cable Board in Suva was an impressive structure with wide sweeping verandahs on three sides. It was ideally situated on the hill overlooking Albert Park and the Grand Pacific Hotel. It was not far from Government House. Originally it housed the large bachelor section of the staff, but when manual working gave way to automatic relays and a consequent reduction in staff, it became redundant. It was then converted into a high-class boarding establishment called the "Oceanic."

What did all those active young men in the staff quarters do when house-bound by the torrential downpours so prevalent in Suva? Some genius dreamed up the chariot race. It was a combination of athletic prowess and imagination that would have created a furore at the Olympic Games!

The first-floor bedrooms opened onto the wide verandah running round three sides of the building. The "track" would be prepared by rubbing the verandah floor liberally with laundry soap to provide a lubricated fast service.

The chariots? Well, every bedroom boasted a washstand equipped with an ewer of water, a large basin and that utilitarian article known as a chamber pot. One is reluctant to become lyrical about chamber pots but they were the essential item in the chariot race. They were described by various informants as "substantial," "massive" and "commodious."

Sheets stripped from the beds were rolled and threaded through the handles of the chariots and, with a bloke pulling each end of the sheet, and a charioteer sitting astride the chamber pot, the team was complete. The teams raced one another round the verandah, the chariots slithering on the soaped floor. One can imagine the pile-ups and crashes as the chariots swung wide at the corners. Needless to say the races relieved boredom and worked off high spirits on many a wet day. All breakages were covered by deductions from salaries at the end of the month.

On one occasion there was a gymkhana in Albert Park with a grandstand full of notables. The program indicated that the staff of the Pacific Cable Board would provide a "novelty event." You guessed it! When the time came they lined up with mysterious parcels under their arms. When unwrapped the parcels disclosed sheets and chamber pots. Sheets were threaded through handles and charioteers seated themselves at the starting line. At the starting gun away went the chariots skidding over the slippery turf to the amusement of the crowd and the embarrassment of the Superintendent and his lady.

CHAPTER
TWENTY-TWO

# Bamfield 1935-1942

I WAS SO GLAD TO GET BACK TO Bamfield that I asked to be allowed to remain there indefinitely. The authorities agreed, subject to the exigencies of the service. I stayed until I retired twenty-five years later.

In order to compensate for the lack of interest in a system that was fast becoming completely automated, I took up outside interests. I began advocating a park on the scenic southwest coast of Vancouver Island. It was to take nearly thirty years of advocacy before my dream of a park on the southwest coast of Vancouver Island became a reality. (It was called the Pacific Rim National Park, and over 500,000 people visit the area each year.)

When World War II was declared in 1939, a detachment of the Canadian Scottish Regiment was dispatched from Victoria to Bamfield to guard the cable station against sabotage. They were temporarily accommodated in the Cable Hall until barracks could be built on one of the concrete tennis courts. The station was encircled by barbed wire and sentries placed at strategic points. Military authorities considered camouflaging the station and sent experts to see how this could be done. However since the station was situated at the junction of two inlets, pinpointing its location, it was decided that camouflage would not serve any useful purpose.

As in World War I, the cables were filled to capacity with vital government information as well as business and social messages. The staff was placed on wartime footing and worked an average of ten hours a day, night and day, every day until the war ended six years later. Once again lives consisted mainly of working, eating and sleeping.

As the station had not yet been mechanized, the "gum and punch" method still prevailed. Reading the long-distance cable signals demanded constant concentration and was a physical as well as a mental strain. There was no time for social occasions. Pensioners were asked to return to work and some did.

On 20th June, 1942, the war came close to home when a Japanese submarine shelled the lighthouse at Estevan Point, about fifty miles farther up the coast. Bamfield was immediately placed under black-out conditions and the guard was doubled. All eyes peered out into the darkened night looking for submarines. It was an eerie feeling to think that, at any moment, a shell might come crashing through the office walls. That was the first, and only enemy action on the west coast of Canada.

Since radio transmissions can be intercepted and decoded by the enemy, all important information is routed via the cables. Cables can be cut by the enemy but they cannot be tapped.

When the Italians entered the war in 1940 they cut all the cables in the Mediterranean Sea. This meant that all urgent traffic from the battlefront in Africa, where the war against Rommel was raging, to the War Office in London, had to circulate via the Indian Ocean, Australia, the Pacific Ocean, Canada and across the Atlantic Ocean to London. The Pacific cables were "red hot" with long urgent "Clear the Line" cipher and code messages which all had to be checked for accuracy.

The well known foreign correspondent, Cecil Brown, sent graphic press dispatches describing the fall of Singapore to newspapers in New York. While being relayed through Bamfield they made grim reading. His final message on 11th February 1942, said:

> Now closing down. Most unlikely able to evacuate. Please inform wives. Goodbye.

That was the beginning of the end of the British Empire.

During the war I got married to an American tourist, Pauline Head, whom I had met while she was on a west coast tour on the *Princess Norah*. As it was wartime it was not possible to obtain leave of absence. All I could get was a weekend off. We were married in Christchurch Cathedral in Vancouver, and spent our honeymoon at the Cameron Lake Chalet near Port Alberni. We travelled by bus and ferry from Vancouver to the Lake, complete with all our wedding presents and were met at the entrance to the long driveway to the Chalet by the proprietor who had a wheelbarrow to transport our luggage. He was somewhat taken aback at the sight and size of our pile of luggage and had to make several trips to move it all.

Spectacular fires lit the night sky above the lake. It was the season for forest slash burning and some of the fires had got out of control.

Upon our return to Bamfield we lived in a small seaside cottage that I owned, there being no accommodation available on the Cable Station.

I carried my bride over the threshold in the traditional manner. I apologized for the smallness of the place, explaining that although I was entitled to a house on the station, there was none available until a vacancy occurred. I then lit the oil lamps and made a fire in the kitchen range.

Married men living on the station had every convenience provided. If they wanted wood or coal, all they had to do was put a memo in the manager's basket and it would be supplied and even stacked.

Living off the station, I had to do everything for myself. There were no tradesmen in Bamfield. One had to be a jack-of-all-trades to maintain a home on the West Coast. As Bamfield was in unorganized territory there were no municipal systems of power, water or sewerage. I had to install and maintain my own. Being wartime, things were hard to get. I did the best I could with what was available, but that wood stove made a slave out of me. Having to work ten hours a day, every day of the year, relaying messages in the office, my "spare" time was spent on the woodpile. There was no truck to dump a load of wood at my back door—it had to be brought around by water in blocks and split on the beach. Then each piece had to be thrown eight times before it was stored in the basement.

But life in Bamfield is another story.

At the beginning of World War I, the British cut and commandeered the German trans-Atlantic cables which ran from Bremen to New York. The operation was a hazardous one since German submarines were very active in that area. HMS *Highflyer* stood by while the cableship worked on the cables. Later she was relieved by a French cruiser.

The northern end of one cable was re-routed to England and handed over to the Pacific Cable Board which now had two cables across the Atlantic. The second one was given to the French and re-located between France and the islands of St. Pierre and Miquelon. Part of one of the cables left over from this operation was relaid between Clover Point, Victoria and Bamfield, via the Strait of Juan de Fuca. It was an emergency route in case the landline between Bamfield and Parksville was interrupted. This old cable was frequently interrupted on the rugged west coast of Vancouver Island and, although never used commercially, the government insisted that it be maintained—especially in time of war.

The only time it was ever used except for tests was when my wife, Pauline, was in a Victoria hospital awaiting the birth of our child, Susan, during World War II. As the time approached, I got on the key and called the operator at the army headquarters in Gordon Head, Victoria where the cable terminated. I asked him if he would telephone the hospital and inquire about my wife's condition and let me know when the child was born. Not long afterwards he called back and signalled, "It's a girl! Both doing well." So, although it had cost the Cable Company tens of thousands of dollars to maintain this cable, the money was not entirely wasted.

CHAPTER
TWENTY-THREE

# Fanning Island and Suva during World War II

AT THE BEGINNING OF WORLD WAR II, Fanning Island was protected by a New Zealand garrison with a six inch gun previously mounted on South Head, at the entrance to Sydney Harbour.

In 1940 a Fanning Island Defence Force was formed to augment the New Zealand garrison, its members being drawn from the Cable Station staff, plus a platoon commanded by a native sergeant. The Force was captained by E. M. Dennis, with A. H. Evans as lieutenant. The two non-commissioned officers were Sergeant Frank McCay and Corporal Maurice Rees. Privates were Norm Giles, Harry Fox, Viv Molyneux, Ted Waldron and Jack Guthrie.

The Force was equipped with .303 army rifles and one Lewis machine gun.

Alarms, which caused a mild panic, were fairly frequent as stray ships and planes were spotted by the army lookouts. Fortunately there was no enemy action.

In 1942 the New Zealand garrison was replaced by several hundred United States servicemen and the need for the Defence Force no longer existed.

As the threat of hostilities with Japan increased, women and children were evacuated to Honolulu. Ironically they arrived off Honolulu in the small vessel *Dickenson* on the morning of 7 December, 1941, just as the Japanese launched a full scale aerial attack on Pearl Harbor.

In 1945, Fred Studman was transferred from Auckland to Suva to augment the staff under war time pressure. Here, as in Auckland, the office was sandbagged and boarded up for protection against air raids.

An account of Studman's voyage to Suva follows:

I travelled on the old *Tofua* carrying twice its normal complement of passengers. Twelve of us slept on mattresses on hard boards in the forward hold. Women were accommodated in tiered bunks in the saloon. The ship zigzagged all the way to Suva and was completely blacked out above decks at night as a precaution against submarines. This was definitely not the first class travel we were accustomed to.

Upon arrival in Suva, I was accommodated in the Pier Hotel for the two years I was there. We were well looked after, meals being brought to the office during long shifts. We had little or no time for social activities.

22

23

22 Norfolk Island Cable Station and Residence, *c.* 1930. AOTC ARCHIVES

23 Norfolk Island circuit siphon recorder, *c.* 1921.

ALL PHOTOS COURTESY OF THE AUTHOR UNLESS OTHERWISE NOTED

24

25

24 Suva Cable Station, *c.* 1924. AOTC ARCHIVES

25 Pacific Cable Board bachelor quarters, Suva, 1913. AOTC ARCHIVES

26

27

26 New Zealand cableship *Iris* in Bamfield Inlet, 1903. Later she was renamed *Recorder* by C. & W. Ltd.

27 Bamfield defence force, World War I, *c.* 1914.

28

29

28 Staff—Bamfield Cable Station, 1937. BACK ROW: Forder, Le Quesne, Forsythe, Salmon, Gallagher, Greenlees, Rockley, S. Denison, Barber and McCormick. FRONT ROW: Scott, Baxendale, Beehan, Woodall, Fraser, Butcher, Bolin, Richards and Fisk.

29 Bamfield Inlet looking west across Aguilar Point to Barkley Sound. CPR's *Princess Norah* at Bamfield east, *c.* 1940. B.C. ARCHIVES AND RECORDS SERVICE 2674

30

31

30 Bamfield Cable Station. Concrete office building in foreground. Bachelor quarters and housing for married couples in background.

31 Canadian-Scottish sentry on roof of cable station office building during World War II.

32

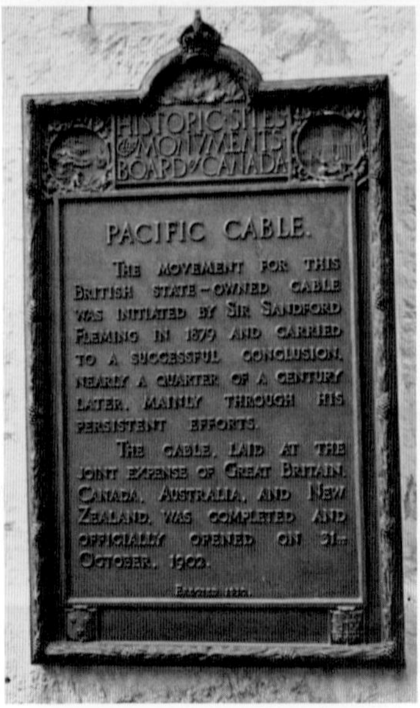


33

34

32 Cable veterans and families return to Bamfield.

33 Old plaque, dedicated in 1931, announcing the completion and official opening of the Pacific cable on October 31, 1902.

34 More recent memorial, marking the end of an era. LEFT TO RIGHT: Earl Pallister and Bruce Scott, July 24, 1977.

35 

 36

35 Friend, fellow cableman and one of those recalling stories for this book, William Duffield Fowlie.

36 Display and description in Parks Canada, Wickaninnish Centre, Pacific Rim National Park, West Coast of Vancouver Island.

ENGLISH TRANSLATION:

*One of the key advocates for the development of a National Park on the west coast of Vancouver Island is R. Bruce Scott.*

*An author and long-time resident of the Barkley Sound region, Scott combed historical archives and interviewed local residents to ensure that the colourful history of the area was preserved.*

*These exhibits depicting west coast Indian culture are a result of the generous donation of Scott's collection of artifacts.*

PHOTO BY SUSAN SCOTT

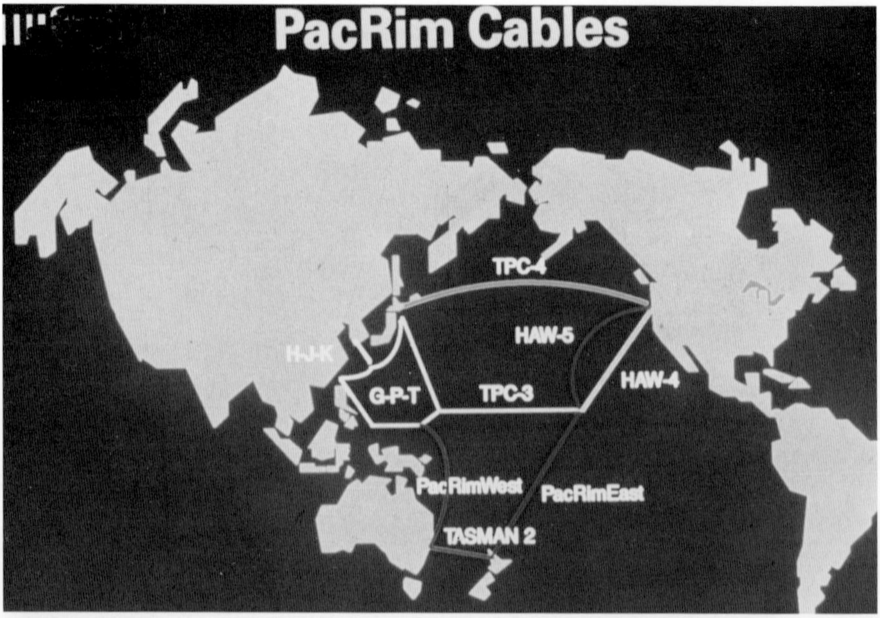


37

38

37 Map showing latter-day layouts of modern, lighter-weight cable in the Pacific Ocean.

38 The French cableship *Vercors* in Auckland harbour just before bringing to shore the TASMAN 2 cable.

CHAPTER
TWENTY-FOUR

# Automation

DURING THE CABLE AND WIRELESS era radical changes were made to the method of operating the Pacific cable. After the war, "gum and punch" operating was superseded by the regenerator system with automatic printers at terminal stations. Reperforators, which received the messages on tape ready for onward transmission, were installed on all circuits at Bamfield.

Synterpolators, when phased correctly, separated the three channels, A, B and C superimposed on the single core of the fast cable. Synchronization was controlled by a master clock governed by a vibrating fork. All stations on the line across the Pacific were synchronized to the transmitting terminal.

From 9 p.m. to 9 a.m., traffic from Australia and New Zealand bound for North America and Europe, would pour in over three channels on the fast cable, each working at 250 letters per minute. "A" channel would contain traffic from Sydney, "B" from Auckland and "C" from Suva. The tape was scrutinized by the operators for failures. If reception was faulty the distant station would be stopped and the failures pointed out to all relay stations on the line. Specimen signals would then be transmitted by the distant station until the fault was localized and rectified.

The operator manning the transmitter on the landline to Montreal would take batches of the received tape from the cable and feed it into the transmitter, giving priority to full rate messages. Usually there would be little or no outbound traffic at that time since it would be in the early hours of the morning in London.

When all homeward traffic was cleared the fast simplex cable would be switched over for outward transmission and the slow cable used for control and homeward traffic—if any.

Traffic would now start pouring in over the trans-Canada landline. The traffic was received on perforated tape ready for onward transmission on A, B and C channels on the fast cable. The landline was operated at a fast speed, the tape being checked by an operator. Should reception be faulty, Montreal would be stopped and the C.P.R. requested to check the line. Should traffic received over the two Atlantic cables begin piling up in Montreal, a second wire across Canada would be requested and supplied by the C.P.R.

There were six operators and a supervisor on each shift. To relieve the tedium, operators changed circuits every hour—from transmitting to receiving and from landline to cable operation. At the end of each shift, no man could leave his post until he was relieved by his successor on the next shift, so any latecomer was most unpopular. Work was always referred to as a duty. One was either on duty or off duty.

It was always a relief when all traffic was cleared from either Sydney or Montreal and the transmitting station would signal "N N CMG N N" which meant clear of all traffic. This was especially welcomed at Christmas time when, after days—even weeks—of continuous transmission night and day, the staff could relax and enjoy the Christmas festivities, with only a minimum of staff on duty. During idle periods, the staff gathered around a pot-bellied stove, made toast and "chewed the rag." Although the office was heated by hot water radiators, the stove provided additional comfort in the way of direct heat.

The transition from "gum and punch" to reperforator attendant had a demoralizing effect on the cable operator. It was not submarine telegraphy any more—it was electronics. No longer was he the linchpin in the service. All he had to do was scrutinize tape automatically received, looking for mechanical failures before it was automatically re-transmitted.

It was very depressing.

CHAPTER
TWENTY-FIVE

# Nationalization

IN 1945 THE AUSTRALIAN government, which had a long-standing policy of nationalizing all public utilities, proposed that each country in the Commonwealth should take over its overseas telecommunications. Under the Anzac Scheme, as it was called, each country would form its own corporation to own and control its overseas telecommunications. Cables, cableships and intermediate relay stations would be owned and maintained by Cable and Wireless Ltd.

The scheme was eventually accepted by all Commonwealth countries. In 1950 the Canadian government took over the Canadian section and formed the Canadian Overseas Telecommunication Corporation, a crown corporation. The Australian government likewise formed the Australian Overseas Telecommunication Commission.

The Canadian government now concentrated on the improvement of its overseas communications and the service became further automated. Relays superseded reperforators. Now signals flashed through Bamfield instantaneously, requiring only one man on duty each shift.

The Corporation's strategy was to eliminate Bamfield entirely. The handwriting was on the wall. Bamfield station was doomed.

In anticipation of the Bamfield closure, I bought a house in Victoria and moved my family there. I was due to proceed on long leave prior to retirement at age fifty-five in November 1959. Not wishing to be transferred to Port Alberni or Montreal for the remaining five months of service, I volunteered to remain behind in Bamfield to wind up affairs and dispose of the assets of the station.

My offer was accepted.

During the previous months, a new cable office had been built in Port

Alberni and a connecting cable laid up the Alberni Inlet to Port Alberni by the cableship *Monarch*. New electronic equipment had been installed and tested, and all was now ready for the change-over from the old system to the new.

It was made on 20th June, 1959.

At the appointed time the few remaining staff at Bamfield, Baxendale, Elliston, Bailey, Gordon, Wallace and Scott, sent the following message to all stations on the line across the Pacific:

> Farewell from Bamfield before closing down after fifty-seven years of operation and leaving it to the shades of the past.

A footnote to Fanning Island, the next relay station, was added:

> The next signals you see will be those of Port Alberni. Goodbye, old man.

The cables were joined together at Bamfield and that was that!

An era had ended!

Three operators were transferred to Port Alberni station and the manager, H. L. Baxendale, and engineer, Nigel Elliston, proceeded on leave prior to retirement. I was left with four Chinese servants and a stationary engineer, H. Wiseman, to run the power plant to provide me with lights.

All the instruments, which throughout the years had attained a remarkable degree of electro-mechanical perfection, were covered with white sheets and left where they were, making the operations room look like a morgue. They were obsolete.

The lights of the office building, which had served as an aid to navigation (being visible at sea) were now extinguished. In their stead the Department of Transport, Marine Division, erected a flashing light on the rocks of Aguilar Point to serve the same purpose. The empty office building, which now seemed like an oversized mausoleum, served only as a cable hut wherein the cables were joined together, bypassing Bamfield.

At that time (1959) this was considered to be a radical change in the method of operation of the cable, but it was nothing compared to the change that was to come within the next few years. Not only did the system of operating become obsolete, but so too did the cables. Telegraphic cables gave way to telephonic ones. The new type of coaxial telephone cable, which had proved so successful across the Atlantic Ocean, was completely different in construction from the old telegraphic type. It was comparatively lightweight and flexible, containing built-in amplifiers every twenty miles, which permitted the transmission of wide frequency bandwidth necessary for the numerous channels of telephonic and telegraphic communications.

## Footnote 1994

Australia Overseas Telecommunication Commission has been recently privatized and is now known as Telstra Corporation Ltd. Sydney, Australia.

CHAPTER
TWENTY-SIX

# Last Days of Bamfield Cable Station

After my family left, I moved into the concrete office building. I had the servants move furniture from an empty staff house into what used to be the manager's rooms. My job was to dispose of the assets—mostly furniture, crockery and utensils, in the houses and other buildings. My footsteps echoed along the empty corridors which once had been filled with activity. Hundreds of people had passed through these halls during the years of operation. They seemed to have left a little bit of themselves behind. Now that all was quiet, they seemed to re-emerge from the shadows.

It was a sad task to preside over the funeral of that once great cable station. People came from up and down the coast as word spread of the sale, for there were some very good bargains. Some of the furniture had already been sold to the staff at prices set by the Corporation, but a large amount still remained. The problem was—how to go about it? Being only a small unorganized hamlet, Bamfield had no newspaper in which to advertise the sale. Furthermore, as the fishing season was still open, most of the local men were away fishing. It could only be advertised by word of mouth and by telephoning the residents.

I decided to throw open a different section every few days. As each article had a C.O.T.C. sticker with an inventory number, I gave all persons attending the sale a sheet of paper and an envelope and asked them to make bids on any article they wanted. The envelopes were then sealed and handed to me for processing. In the evenings I went through the stack of bids in order to get the highest bid on each article. The bids were surprisingly low—much lower than the staff had to pay for similar articles. For instance, towards the end of the sale, bids for a solid oak dining table, which could be extended to eight feet, were only $1.00!

Nowadays they are worth close to $1,000! No doubt the canny locals knew that I did not have any choice in the matter. Within a few weeks, every piece of furniture was sold. The buildings had been stripped of their contents.

Now that my work was done, I had nothing else to do but kill time for several months until my retirement fell due. Living in the concrete office building was like living in a mausoleum. The faithful Chinese servants, some of whom had been on the station for over forty years, served me well. I had one to do my laundry, one for housework and another to cook my meals. I was literally waited on hand and foot. The cook caught fish and crabs and served me Crab Foo Yong. Knowing that I liked pies, he picked raspberries, loganberries and apples from the deserted gardens and made pie after pie until I had to cry enough! I dined alone at a long table in the large dining room of the Bachelor Quarters overlooking beautiful Barkley Sound. Its vast emptiness accentuated the sadness of the situation.

As the date of my retirement approached, the servants and the stationary engineer were released. The Chinese were reluctant to leave having spent so much of their lives on the station.

Handing over what was left of the station to an engineer from Montreal head office, I boarded a float plane for Port Alberni and retirement—leaving without a word of goodbye from management after forty years of service.

The seaplane flew via Ucluelet, on the opposite side of Barkley Sound, and then dropped down to land on the water at a lonely float-camp in sheltered waters somewhere near Pipestem Inlet. A lone Chinaman was waiting for the plane. Apparently the camp was being closed down for the winter. Surplus supplies of canned foodstuffs and flour were loaded on the plane, then the Chinaman prepared to board.

The young pilot, noticing a Dalmatian dog nervously slinking around the float, asked him if he was not going to take the dog too.

"Waffor?" he said, adding, "Not my dog."

"But he will die if you leave him behind," said the pilot.

The Oriental just shrugged his shoulders.

"Who does he belong to?" asked the pilot.

The Chinaman named some member of the logging crew, and added: "He no want 'em. He say give him away, but no one to give him to."

"I'll take him then," said the pilot. "You just can't leave him there. Call him over and we'll put him aboard."

The dog nervously approached when called, but backed away when the Chinaman tried to catch hold of him. He played "catch-me-if-you-can" for several minutes before he was finally cornered and heaved and pushed up the iron ladder leading to the cockpit of the plane. There was no place for him to sit except on top of the pile of canned goods and flour bags, which swayed with the motion of the plane. Having no collar I held him by the scruff of the neck so that he would not fall off. It ended up with his sitting on my lap, excitedly peering out the window at the treetops below. The pilot kept looking around to see how the dog was faring. I could see that it was a case of love at first sight. When we landed in Port Alberni, he scrounged a piece of rope for a leash.

"I promised my small son a puppy for Christmas," he said self-consciously. "He'll be surprised when he sees this."

"You have got yourself a good dog," I said. "You won't recognize him when he has had a bath."

I watched them as they headed up Argyle Street for home—a dirty dog on a rope leash, headed for a completely new existence. Then I too headed for a new home and a new life in the city. As the bus pulled out of Port Alberni, I wondered if I would ever see the West Coast of the island again.

A month later, early in December, I was booked to sail on the SS *Orcades* with my wife Pauline and daughter Susan, now fifteen years old, for Sydney, Australia. I was looking forward to seeing my mother again after an absence of twenty-five years. She had sold the family home and was living with my sister Patricia and her husband. Sadly, she passed away a few weeks before I sailed.

For six years the old cable station buildings lay idle. Heating and maintenance of the historic Bachelor Quarters were discontinued. Several offers to purchase the property were made, but none was satisfactory. The Crown Corporation responsible for the property did not seem to know what to do with it. It was offered to the Bamfield Community for a very reasonable price but as Bamfield was unorganized and had no legal authority to conduct the deal, the provincial government would not sanction any transaction.

In 1965, the Bachelors' Quarters, still in a good state of preservation, was demolished—ignominiously pushed over by a bulldozer! Twelve houses, previously accommodating married staff, were put to the torch and destroyed.

Visiting Bamfield at the time, I could not bear to watch the destruction of that beautiful building, in which I had lived for twelve years as a bachelor, but observed it in passing out of the corner of my eye. The original clockwork instruments, used when the station first opened, apparently considered valueless by the authorities, were disposed of to junk dealers. Fortunately their historical worth was appreciated by the Victoria branch of the Institute of Electrical and Electronic Engineers, which in January 1969, honoured the old Cable Station by assembling a permanent display of early submarine telegraphic instruments in the Maritime Museum of British Columbia, some of the old equipment being re-purchased from the junk dealers for that purpose.

In 1969 the cable station property was purchased by the five western Canadian universities of Victoria, British Columbia, Simon Fraser, Calgary and Alberta for a marine biological station, for which it was eminently suited. But millions of dollars had to be spent on new buildings to provide accommodation for the staff and students. The old Bachelors' Quarters, with its more than fifty rooms, would have served admirably for that purpose, had it not been for its seemingly imprudent demolition.

But in the minds of those who had lived in it, the old Bachelors' Quarters still exists—indestructibly delineated in dreams and memories.

CHAPTER
TWENTY-SEVEN

# Relay Stations Closed

In 1962 preparations were made on Fanning Island for the landing of the new trans-Pacific telephone cable, COMPAC. However, partly owing to the indefensible nature of the island, and partly due to the fact that the island was no longer British territory, having recently been included in the new Pacific nation of Kiribati, it was decided to land the cable in Hawaii.

With the opening of the COMPAC cable in 1963, Fanning Island cable station was abandoned and offered for sale.

In 1964 the Fanning Island cable station was taken over by the Hawaiian Oceanographic Institute as a Pacific Equatorial Research Laboratory for the study of equatorial currents, etc., in much the same way as the Bamfield cable station was taken over by the five western Canadian universities as a Marine Biological Station.

The two four-thousand-mile submarine telegraph cables that linked Fanning Island with Bamfield, which for six decades carried the gamut of human emotions, were abandoned. They lie on the bed of the ocean, grounded and bypassed by technological progress. Fanning Island is now known by its original native name of Tabuaeran (Isle of Heavenly Footprint) and is included in the new Pacific nation of Kiribati, which also includes those places previously known as Gilbert Islands, Line Islands and Phoenix Islands.

Sometime in the 1960s, the *Doris Crane*, the sailing ship that had served Fanning Island for so many years, caught fire in the North Pacific and burned to the water line.

In 1962 the cable station on Norfolk Island also became redundant when the new cable, COMPAC, was laid across the Pacific Ocean bypassing Southport, Norfolk and Fanning Islands. In 1972 a radio-telephone station was installed at Burnt Pine, enabling residents to keep in touch with the outside world.

Norfolk Island is a popular holiday resort for people from Australia and New Zealand. A limit has been placed on the number of people who may become residents, the idea being to keep the island primarily for the descendants of the mutineers on the *Bounty* and old-time residents. When William Fowlie revisited the island in 1977, he asked if his name could be placed on the waiting list to become a resident of the island. He was regretfully informed that it would never come to the surface in his lifetime.

When the cableship *Monarch* laid the COMPAC cable in 1963, she bypassed Fanning Island and landed the cable in Honolulu where the cable station was constructed deep in the bowels of a mountain, impregnable to enemy attack. She also bypassed Bamfield, laying the cable up the Alberni Inlet. At the same time, the overhead landline between Port Alberni and Parksville, on the east coast of Vancouver Island, was replaced by an entrenched cable. The project was supervised by Keith Sparks, branch engineer of the Vancouver cable office.

On December 3, 1963, the new telephone cable was officially declared open by Queen Elizabeth. The first stage across the Atlantic Ocean had been completed in 1961; the second stage, Canada to Australia, was completed in 1963, the two stages being linked across Canada by microwave. The new system contained eighty voice channels, each of which could be split up into twenty-two telegraphic channels if required. The system was a duplex one—meaning it could be operated in both directions at the same time. As the cable landed on foreign territory, the system could no longer be referred to as the "All Red Route."

CHAPTER
TWENTY-EIGHT

# Modern Submarine Cables

THE OLD TELEGRAPHIC CABLES OF the Pacific Cable Board era were superseded in 1963 by COMPAC, a co-axial telephone cable containing built-in vacuum tube amplifiers every twenty miles, permitting the transmission of wide frequency bandwidth necessary for numerous channels of telephonic and telegraphic communications. It had an initial capacity of eighty two-way speech circuits, later increased by a system making use of gaps in telephonic conversations which doubled the number of phone calls carried by the system at any one time. However the vacuum tubes had a limited guaranteed design life of twenty years, so a new trans-Pacific cable known as ANZCAN was planned to replace it.

The ANZCAN cable, which cost $400 million, was laid in 1984. It was one of the world's biggest telecommunications projects. The eight thousand nautical mile cable contained 1,123 submersible solid state repeaters—one every six miles. The route, surveyed by HMNZS *Monowai*, a hydrographic survey ship of the New Zealand Navy, was: Australia-Norfolk Island-Fiji-Hawaii-Canada with another segment from Norfolk Island to New Zealand.

The new cable contained 1,380 circuits, more than seventeen times the capacity of COMPAC, using telephonic, telex, facsimile and data channels. Its total length is approximately 8,175 nautical miles, a nautical mile being equal to one and one sixth of a mile. It has a guaranteed life of twenty-five years, and is still in operation.

Cable and Wireless cableships, *Cable Venture* and *Mercury* were chartered to lay the various segments of cable. The first lay, between Canada and Hawaii, was made by *Cable Venture* in November 1982, the remainder being completed in 1984. New cable stations were built in Norfolk Island, Hawaii and Port Alberni.

Just as COMPAC cable was superseded by ANZCAN, ANZCAN will soon be superseded by radically different optical fibre cables, which will revolutionize telecommunications. Optical fibres transmit messages from point to point as a succession of laser generated light pulses through hair-thin strands of glass. Using light instead of electrical signals they will offer more than thirty times the capacity of the cables they replace.

The optical fibre cable is only 26.2 mm in diameter, the outside being composed of tough plastic. The glass fibres in the core are variously coloured to simplify identification during splicing. They contain two pairs of fibres—one for traffic and one spare—which are covered by several forms of protection and a braided copper sheathing that carries the high voltage to power the repeater amplifiers.

A network of high-quality optical fibre cables is planned for the Pacific region. When complete in 1994, the South Pacific network will extend more than 16,500 kilometres across the Pacific Ocean. It will consist of three submarine optical fibre cables, TASMAN 2, PACRIMEAST and PACRIMWEST.

TASMAN 2 will link Australia and New Zealand. PACRIMEAST will link New Zealand and Hawaii (bypassing Suva) with connections to North America and onwards to Europe. PACRIMWEST will link Australia and Guam with onward connections to Japan and Asia. TASMAN 2 will also link PACRIMEAST and PACRIMWEST forming a unique telecommunications loop in the Pacific providing alternative routes in the event of interruption.

TASMAN 2 was laid between Sydney and Auckland in 1991 by the French cableship *Vercors*. It is 2230 km long and cost US $110 million. It contains 19 built-in repeaters spaced 116 km apart, and has a capacity of 60,000 voice circuits.

The northern section of PACRIMEAST was laid in 1992 from Hawaii to a point 60 km east of American Samoa, a distance of 5350 km, where the end was buoyed. The southern section was then laid from Auckland to the same point, a distance of 2500 km, by CS *Vercors*. The two sections were then spliced together and lowered to the bottom of the ocean.

Containing 56 repeaters, it has a capacity of 30,000 voice circuits. Its total length is an incredible 7850 km. It was laid at a maximum depth of 7500 metres at a cost of US $270 million. It is now in full operation.

PACRIMWEST between Sydney and the island of Guam, will be laid in 1994. Its length of 6980 km will contain 53 repeaters spaced 130 km apart; its capacity will be 30,000 voice circuits and its cost, US $270 million. Maximum depth will be 8500 metres. The total cost of the whole network will be US $650 million.

Since optical fibre cables use light instead of electricity for the transmission of information, they are not affected by electromagnetic, atmospheric or radio interference. Nor do they have to contend with electrical resistance and capacitance which limited the speed of the original trans-Pacific cable between Bamfield and Fanning Island, to only one dot in 9.3 seconds. Instead the laser light impulses travel at the speed of light: 299,330 km (186,000 miles) per second.

The built-in repeaters regenerate the signals overcoming losses due to distance; they are powered by 7,000 volts of electricity applied by a secondary copper sheath in the cable. To guard against damage by ships' anchors in coastal waters, the cable has been entrenched, and in deeper waters, protected from damage by predatory sharks by using armoured cable.

The system has a predicted service life of twenty-five years.

Modern cableships, painted white like cruise ships, have graceful, yacht-like lines. Equipped with the latest high-tech computerized equipment, the bridge is an expensive collection of computer terminals and display screens. The latest satellite navigation system enables them to pin-point positions in the vast expanse of ocean within a few metres.

To service this and other networks of submarine cable, Cable and Wireless (Marine) Ltd. maintains a fleet of cableships located at strategic places around the world, none more than ten days' sail from the most remote section of cable. They are kept fully commissioned, prepared to sail within twenty-four hours.

In April, 1991, two successive breaks occurred off Bondi Beach in Sydney, Australia. The first was caused by an unidentified deep-sea trawler fishing for orange roughy. The cableship *Pacific Guardian*, based in Suva, Fiji, repaired the fault and was on her way back to Suva when she was diverted back to repair the second fault.

When the second fault occurred, tests located it 50 km off Bondi Beach, the same location as the previous fault. A helicopter was chartered immediately and flown out to the location of the fault where a deep-sea trawler was positively identified. Attempts to contact the vessel by radio were unsuccessful so an all-port bulletin was issued requesting that O.T.C. be advised when the trawler entered port. Next day information was received that the vessel had entered port in Sydney to unload its catch. A search warrant was obtained and executed by the federal police. The ship's log showed that it had been trawling near the cable fault at the time of the failure. Examination of the trawler's otter boards revealed minute tracings of polythene and aluminum from the ANZCAN cable and the Scientific Squad was called to document the evidence.

Cable companies will gladly compensate owners of ships for the loss of their anchor and chain if reasonable proof is produced that they were in the vicinity of a submarine cable, and that they hooked the cable with their anchor and had to cut it loose rather than run the risk of damaging the cable by trying to clear it with brute force. However, it is illegal to anchor over charted cable routes in coastal waters.

From Hawaii, PACRIMEAST is linked to San Francisco and Vancouver by North Pacific cables HAW-4, HAW-5 and TPC-4, thence across the continent by optical fibre channels provided by Teleglobe Canada, BC Telephone Co., and other North American carriers, thence on to Europe by trans-Atlantic optical fibre cable, TAT-9.

The first section of the North Pacific system (TPC-4) from Port Alberni to a point midway between San Francisco and Port Alberni, was laid by the cableship *Global Sentinel* in July 1992 and buoyed. The second section from San Francisco to this point was then laid and connected to a branching repeater. Then the third section was laid from there to Japan, a total distance of 8500 km. The cable contains three sets of optical fibres and sixty-seven repeaters powered by 6,500 volts at 1.6 amps from each end. It has a capacity of 15,120 channels which can be increased up to five times with the addition of special equipment.

Besides TPC-4, there is also NPC-1 connecting Japan to Alaska and San Francisco, TPC-2 connecting Hawaii to Guam and on to Asia, and TPC-3 from Hawaii to Japan and Guam. The ANZCAN cable is still in service.

Most of the optical fibre cable in the PACRIM system was manufactured by the Tasman Cable Company's plant in Botany Bay, near Sydney. Part of PACRIMEAST was supplied by AT&T Submarine Cable Systems of USA.

Half of the North Pacific system cable was built by AT&T, the other half by KDD (Japan).

Nowadays most businesses have fax machines for communicating with other local and overseas firms. The machines are connected via telephone lines to cable terminals. Each firm has a fax number, like a telephone number. To send messages to another firm overseas, the distant fax number is dialled and a written or typed message or document is transmitted. It will be stored in a computer and sent when the line is clear to the distant firm, being received in printed form on the fax machine at any hour of the day or night.

After their original transmission from terminal stations, messages are not touched by hand again. They flash through relay stations and are automatically printed at their destinations.

It is a long way from the old days of "gum and punch" human relay operating!

CHAPTER
TWENTY-NINE

# Veteran Cablemen

ALL CABLE COMPANIES HAVE Veteran Cablemen's Associations which hold annual reunions to renew friendships and discuss the "good old days" spent at various isolated and city stations. The twenty-fifth anniversary of Teleglobe Canada's Quarter Century Club, Western Canada Division, was held in the Empress Hotel in Victoria in April, 1982. I attended, accompanied by my wife, Pauline. It was a large gathering of over one hundred people. First there was a business meeting for the men who then joined the ladies for cocktails before dinner.

Just as the soup was being served, William Fowlie, who had just returned that morning from a trip to Australia, arrived to loud acclaim. When his turn at the microphone came, he spoke about Fanning Island days and then referred to Auckland where he had met me, saying that he remembered that I used to wear a pin-striped suit and a polka dot tie. We had not met since.

After dinner I made my way over to him and invited him to stay with me at my seaside cottage in Bamfield for a week. He said he would be delighted as he had many happy memories of Bamfield.

The cottage was perched high atop a hundred foot cliff and commanded a view of Barkley Sound and a glimpse of the Pacific Ocean. It was nearing the end of the herring fishing season and the Sound was dotted with fish-boats by day. At night the red, green and white navigation lights of the numerous vessels twinkled like those of a city. The partially sheltered waters of the bay below were creamy with herring spawn, and with the spawn came thousands of sea-birds and scores of sea lions, all competing for the spawn.

During his stay, Fowlie hiked over the lifesaving trail to Pachena

Beach, five miles away. "I had to revisit the place," he said. "One night, in my courting days, my wife and I danced on the sand there, in the moonlight."

One evening I gave him a brief copy of my reminiscences, which included the time in Auckland when I had a temporary breakdown due to mechanical failures. After reading them he became unusually quiet, so I asked him what he thought of them.

It was at this moment that he made his confession to me. "I was responsible for your breakdown in Auckland because it was I who carelessly misadjusted that Kleinschmidt perforator. I am sorry we did not realize it the first time it happened."

So, over half a century later, I received a belated apology, and thanked him for letting me know.

A few days after our arrival in Bamfield, a howling southeaster blew, and rain poured down in typical west coast fashion. It was too wet to go outside so I thought it would be a good opportunity to talk with Fowlie about his experiences at various cable stations in the Pacific. I set up my tape recorder and began questioning him. The information I gathered while the storm raged about us, provided much of the historical drama in this book.

CHAPTER
THIRTY

# Old Cable Saves Ship

One day in October 1991, the Canadian government fisheries patrol vessel, *Tanu,* entered Bamfield Inlet just in front of the old cable station. As she approached the dock on West Bamfield where she intended to moor, her captain reversed engines in order to slow her approach. Instead the engines speeded up uncontrollably in reverse. The ship was headed straight for the rocks at the entrance to nearby Grappler Inlet. The bow anchor was dropped but did not hold. Instead it just dragged along the bottom. The only way the captain could stop the engines was to cut power to the ship, which also cut power to the rudder. Still under way, there was nothing the captain could do to prevent the ship from crashing onto the rocks on the north side of Grappler Inlet.

Suddenly the ship was brought to an abrupt stop. The anchor had hooked the heavily armoured shore-end of one of the abandoned trans-Pacific cables in front of the cable station—much to the relief of the captain and crew.

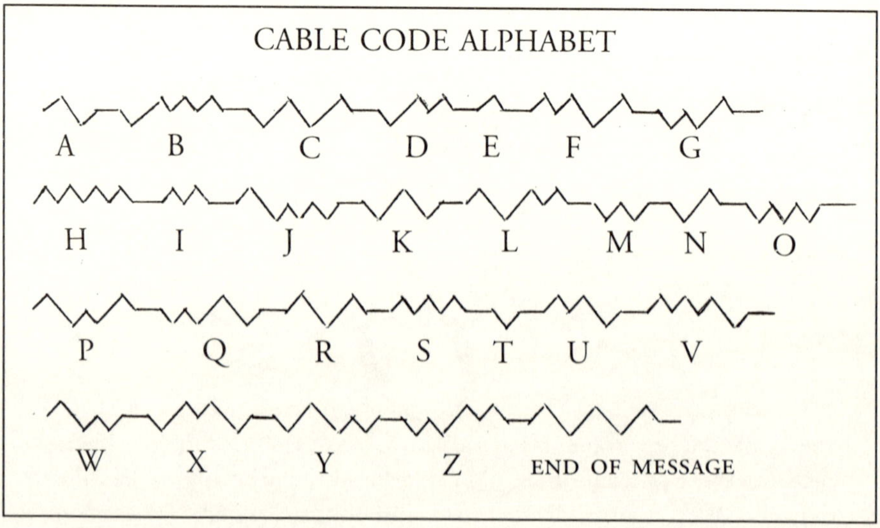


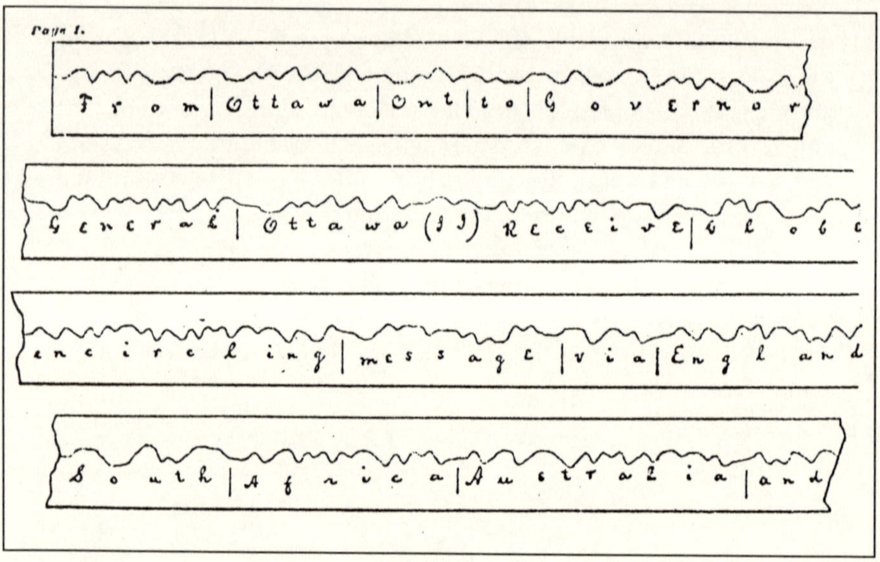


The first world-circling cable coded message from Sir Sandford Fleming to His Excellency Lord Minto.

See delivered copy of this telegraph on page 126.

Page 2.

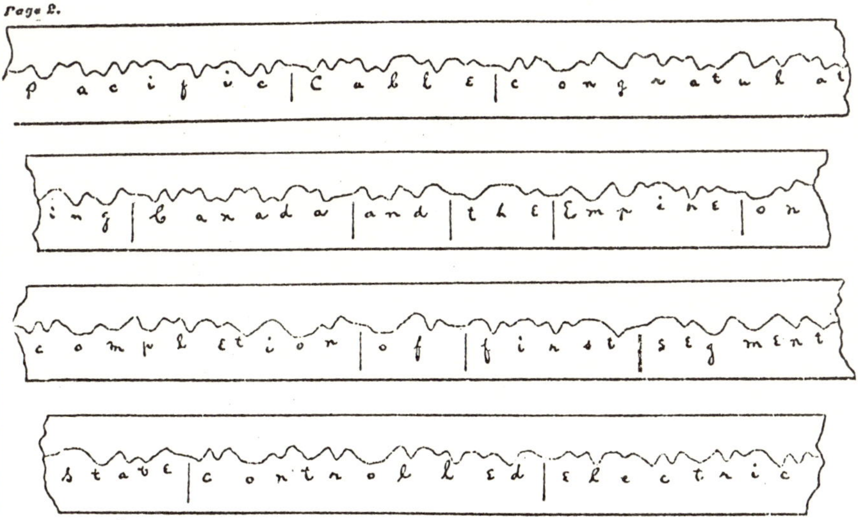
pacific | Cable | congratulat
ing | Canada | and | the | Empire | on
completion | of | first | segment
state | controlled | electric

Page 3.

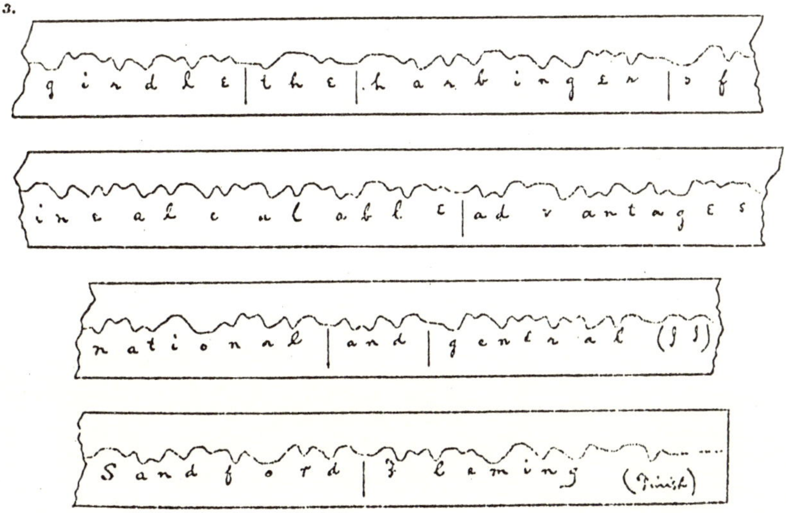
girdle | the | harbinger | of
incalculable | advantages
national | and | general (gd)
Sandford | Fleming (Finish)

FIRST WORLD-CIRCLING TELEGRAPH MESSAGE

The message delivered to His Excellency Lord Minto at Government House, Ottawa, Nov. 1st, 1902, after circling the globe.

---

Canadian Pacific Railway Company's Telegraph

All messages taken by this Company are subject to the conditions printed on our Form 2.

10. RA. MD. RO. (39 Words) 8.35 A.M.

(via Commercial Eastern Australia Pacific)

To Governor General,
Ottawa.

Receive globe encircling message via England, South Africa, Australia and Pacific Cable congratulating Canada and the Empire on completion of the first segment state controlled electric girdle the harbinger of incalculable advantages, national and general.

From Ottawa, Oct. 31st, 1902 Sandford Fleming.

---

The Pacific Cable Board.
Bamfield B.C. Station.
Nov 18th 1902.

Dear Sir,

I enclose slip of your "Round the World cable". as we do not require its return. I shall be pleased if you will retain it as a memento. I have marked it in plain characters so that you will be able to follow it. The characters in brackets (II) and (finish) represent the division between address & text, text and signature and completion of cable respectively.

I am
Yours faithfully
A G McLachlan

Sir Sandford Fleming K.C.M.G.
Ottawa

# Bibliography


CHAPTER 3 A.O.T.C. *Transit* magazine, January/February 1991, April 1992

Hugh Barty-King, *Girdle Round the Earth*, Heinemann London 1979

CHAPTER 4 Hugh Barty-King, *Girdle Round the Earth*

W. P. Daykin, Carmanah Point Lighthouse diaries

Provincial Archives of British Columbia

*Colonist* Newspapers, Victoria, B.C., 1901-1902, 1926

CHAPTER 5 *Colonist* Newspapers, Victoria, B.C., 1902

CHAPTER 7 A.O.T.C. *Transit* magazine, February 1978

W. D. Fowlie Interviews by R. B. Scott

Provincial Archives of British Columbia

CHAPTER 8 W. D. Fowlie Interviews by R. B. Scott

Miles Smeeton, *The Sea Was Our Village*, Gray's Publishing, Victoria, B.C., 1973

CHAPTER 9 C. B. Whaley, *Chateau on the Bluff*

*The Bamfielder*, August, September, October 1977

CHAPTER 10 W. D. Fowlie Interviews by R. B. Scott

Alberni *Pioneer News*, 5 March 1910

CHAPTER 11 Robert Marchant, unpublished

CHAPTER 14 W. D. Fowlie Interviews by R. B. Scott

CHAPTER 15 W. D. Fowlie Interviews by R. B. Scott

CHAPTER 16 W. D. Fowlie Interviews by R. B. Scott

CHAPTER 17 W. D. Fowlie Interviews by R. B. Scott

CHAPTER 18 *Port Alberni News*, 21 September 1912

*Port Alberni News*, 20 June 1914

CHAPTER 19 Hugh Barty-King, *Girdle Round the Earth*

CHAPTER 21 Fred Studman, Auckland, New Zealand, unpublished

John Lee, Auckland, New Zealand, unpublished

CHAPTER 22 R. B. Scott, *Bamfield Years: Recollections*, Sono Nis Press, Victoria, B.C., 1986.

CHAPTER 23 R. H. Payne, *Farewell Fanning Island*

*Transit* magazine, November 1963, February 1964

Jack Guthrie, F.I. Defence Corp

O.T.V.A. Newsletter

Fred Studman, A.O.T.C. Newsletter, November 1990

CHAPTER 28 A.O.T.C. *Contact*, March 1981

A.O.T.C. *Transit* magazine, January/February 1983, June 1991

A.O.T.C. South Pacific Network brochure

# *Index*